AF342245

GEOP
Gas Engineering and Operating Practices

A Series by the Operating Section
The American Gas Association

Volume I

SUPPLY

Book S-1
Underground Storage

The American Gas Association
Arlington, Virginia

Library of Congress Catalog Card Number: 85-70460

ISBN 0-87257-006-1

A.G.A. Catalog Number XY9008
1M12.90-138-69   Printed in the United States of America

CONTENTS

PART 1 Underground Storage of Natural Gas

PART 2. CASE STUDIES

CHAPTER 5. DEPLETED HIGH-PERMEABILITY
               GAS RESERVOIRS

List of Tables and Figures

FIGURES

TABLES

PREFACE

Arlington, Virginia
December, 1990

Underground Storage is the seventh of 11 books that will constitute the six-volume Gas Engineering and Operating Practices (GEOP) series of texts addressing the various technical aspects of gas supply, transmission, distribution, measurement, utilization and related technical services subjects.

The other book in the Supply Volume, *Supplemental Gases—Peak Shaving/Base Load,* treated one major method of providing adequate gas supply during periods of peak demand. Where it is available, underground storage of natural gas near gas markets, has long been the preferred method of supplying the extra gas needed to meet peak gas demand because of its sheer volume and its economic and technical advantages over supplemental gases and other forms of storage.

In 1989, 395 storage pools in 27 states offered an estimated storage capacity of 7.737 trillion cubic feet. This compares to 303 pools in 24 states with estimated capacity of 4.412 trillion cubic feet in 1966. In recent years, peak-day sendout from underground storage has exceeded the amount provided by flowing pipeline gas.

Part I of this book offers a general overview of procedures. Part II provides seven specific examples of how general procedures were adapted for storage in different types of reservoirs.

The GEOP texts represent the writing efforts of hundreds of expert authors from more than 20 A.G.A. Operating Section committees composed of industry consultants, suppliers, and other specialists, in addition to leading pipeline and utility engineers. In the case of *Underground Storage,* the authors wish to acknowledge their indebtedness to the late Prof. Donald L. Katz of the University of Michigan, some of whose many works on the subject are listed in the bibliographic citations at the end of this book.

Readers owe a special debt of gratitude to Anne C. Roess, Librarian, The Peoples Gas Light and Coke Company of Chicago, and her

xiii

colleagues on the indexing subcommittee of the A.G.A. Library Services Committee, who verified references and indexed the book.

We trust that this and the other books in the GEOP Series will prove to be a valuable addition to your professional library.

Gerald G. Wilson
Chairman, GEOP Task Group

Robert L. Parker
Editor

ACKNOWLEDGEMENTS

Special thanks to Transcontinental Gas Pipe Line Corporation: Robert M. Detamore, General Manager, Transmission Services (Retired); Robert G. Mason, Senior Storage Engineer; and Robert G. Graves, Senior Vice President, Transmission, without whose efforts this book would remain unfinished.

CONTRIBUTORS

George E. Babashoff, Associate Engineer, Southern California Gas Company

Kerry J. Comeaux, Director, System Flows, United Gas Pipe Line Company

V. Ralph Cole, Vice President, Valero Energy Corporation

James W. Gourley III, Division Manager (then Manager, Underground Storage), Southern California Gas Company

W. Kenneth Hooker, Gas Supply Manager, Western Gas Supply Company

Vernon R. Hughes, Senior Engineer, Storage, Transcontinental Gas Pipe Line Corporation

Larry D. Krohmer, Resident Reservoir Engineer, Southern California Gas Company

Ralph E. Laenger, Superintendent, Field Service Operations, United Gas Pipe Line Company

William A. McEachin, Storage Rights Administrator (Retired), Southern California Gas Company

Philip S. Magruder, Jr., Executive Assistant (Retired) (then Manager, Underground Storage), Southern California Gas Company

Marvin E. Melton, Senior Petroleum Engineer, Southern California Gas Company

Ali Razavi, Technical Superintendent, Southern California Gas Company

Walter M. Rzepczynski, Manager, Storage Reservoir Engineering, Natural Gas Pipeline Company of America

George E. Stovall, Chief Storage Engineer (Retired), Transcontinental Gas Pipe Line Corporation

William E. Taylor, Bistineau Plant Superintendent, United Gas Pipe Line Company

Steve Webb, District Manager, Valero Transmission, L.P.

Earl L. Welton, Assistant Vice President (then General Superintendent), National Fuel Gas Corporation

Michael J. Whims, Director, Operations, Engineering and Construction, ANR Storage Company

REVIEWERS

Charles H. Becker, Manager, Engineering and Construction, Western Gas Supply Company

John P. Erickson, Staff Vice President, Engineering Services, American Gas Association

Stephen E. Foh, Associate Director, Natural Gas Production and Storage Research, Institute of Gas Technology

Frank P. Hebert, Manager, Onshore Reserves and Gas Storage, Southern Natural Gas Company

H.L. "Tex" Ketcham, Superintendent, Natural Gas Storage and Production, Washington Gas Light Company

Benjamin F. Jones, Manager, Transmission Divisions, Southern California Gas Company

Laurie L. Langer, Manager, Reservoir Engineering, Equitrans

E. Vincent Martinson, Storage Consultant, Northern Natural Gas Company

Richard D. Phillips, Manager, Storage Operations, Southern California Gas Company

Wendell L. Thaler, Vice President, Engineering, Indiana Gas Company

Michael J. Whims, Director, Operations, Engineering and Construction, ANR Storage Company

REVIEW AND INTEGRATION

Reviewing and integrating the writings of the authors into the finished product was the work of still others in the various capacities listed below:

COGNIZANT GEOP TASK GROUP MEMBER

Wendell L. Thaler, Vice President, Engineering, Indiana Gas Company

TECHNICAL REVIEW EDITOR

Stephen E. Foh, Associate Director, Natural Gas Production and Storage Research, Institute of Gas Technology

TEXT REVIEW

Coordination of review or review of manuscript text was performed by the GEOP Task Group of the Underground Storage Committee composed of:

Frank P. Hebert, Manager, Onshore Reserves and Gas Storage, Southern Natural Gas Company

Michael J. Whims, Director, Operations, Engineering and Construction, ANR Storage Company

INDEXING

Members of the A.G.A. Library Services Committee produced the Index under the supervision of **Anne C. Roess**, Librarian, The Peoples Gas Light and Coke Company.

GEOP TASK GROUPS

The creation of the GEOP series began with the establishment of a task group of the A.G.A. Operating Section's Distribution Design and Development (now Distribution Engineering) Committee at the request of the Managing Committee in 1975. The recommendations of this task group were largely followed by the Managing Committee when it initiated the GEOP effort.

The task group members were:

Walter F. Bohls, then Vice President and Chief Engineer, Southern Union Gas Company

Ronald K. Espe, then Vice President, Operations, Pennsylvania Gas and Water Company

Bill E. Hunt, then Supervisor of Gas Planning and Design (Deceased), Illinois Power Company

Emile T. Kaler, Jr., then Manager, Gas Division Engineering Department, New Orleans Public Service Company

Joseph C. Warnock, then Gas Design Section Engineer (Retired), Central Hudson Gas and Electric Corporation

Gerald G. Wilson, Director, Industrial Education, Institute of Gas Technology. Chairman

The members of a subsequent Managing Committee task group that was established in the fall of 1977 and made the final recommendations on how the GEOP Series should be organized were:

Edward J. Filiatrault, Executive Vice President, Northern Illinois Gas Company

Robert H. Holder, then Vice President and Chief Engineer (Retired), Wisconsin Gas Company, Chairman

Henry R. Meyers, then Senior Vice President (Retired), Gas Service Company

Jack N. White, then Vice President, Transmission (Deceased), Panhandle Eastern Pipeline Company

* * *

The following Operating Section members have served on the Gas Engineering and Operating Practices Task Group of the Section Managing Committee through the development of the GEOP Series.

Gerald G. Wilson, Chairman (1989–present), Director Industrial Education, Institute of Gas Technology.

Robert H. Holder, Chairman (1985–1989), Vice President and Chief Engineer (Retired), Wisconsin Gas Company

Anthony A. Bobelis, Chairman (1982–1985) Principal Engineer, Engineering Department, The Brooklyn Union Gas Company

Ty S. Miller, Chairman (1979–1982), Manager, Distribution (Deceased), Southern California Gas Company

C. William Ade (1980–1983), then Vice President, Gas Transmission and Planning, Mississippi River Transmission Company

Anthony A. Bobelis (1979–1987), Principal Engineer, Engineering Department, The Brooklyn Union Gas Company

Barney Champion (1990–present), Vice President, Construction and Gas Supply, Alabama Gas Corporation

William L. Clayton (1988–present), Senior Vice President, Engineering, Operations, and Data Processing, ENTEX, a Division of ARKLA, Inc.

Alton T. Davis (1979–1980), Vice President, Gas Division, San Diego Gas and Electric Company

James W. Garrett (1986–present), Vice President, Operations, Oklahoma Natural Gas Company, a Division of ONEOK, Inc.

Robert H. Holder (1985–1989), Vice President and Chief Engineer (Retired), Wisconsin Gas Company

George M. Hugh (1979–1983), Senior Vice President, Engineering and Operations, TransCanada PipeLines Ltd.

John L. McIsaac (1988–1990), Senior Vice President, Operations, Wisconsin Gas Company

Vernon E. Percell (1983–1988), Senior Vice President (Retired), Gas Operations, KPL Gas Service

Marvin D. Ringler (1979–1981), Engineer, Gas Supply Planning, Consolidated Edison Company of New York, Inc.

O. L. Slaughter (1983–1985), Senior Vice President (Retired), Gas Company of New Mexico

Robert D. Stegner (1983–1989), Senior Vice President, Operations and Engineering, and Chief Operating Officer, Indiana Gas Company

Andrew Tarapchack (1979–1984), Vice President, Distribution (Retired), Washington (D.C.) Gas Light Company

Wendell L. Thaler (1988–present), Vice President, Engineering, Indiana Gas Company

Milton M. Walther (1979–1983), General Manager, Gas Department, New Orleans Public Service, Inc.

Gerald G. Wilson (1989–present), Director, Industrial Education, Institute of Gas Technology

Robert L. Parker (1980–present), Manager, Engineering Series Publications, American Gas Association

OPERATING SECTION
AMERICAN GAS ASSOCIATION

OFFICERS

James W. Garrett
Chairman
Oklahoma Natural Gas
    Company
Tulsa, Oklahoma

William F. Eckles
First Vice Chairman
New Jersey Natural Gas
    Company
Wall, New Jersey

Frank T. Bahniuk
Second Vice Chairman
Elizabethtown Gas Company
Union, New Jersey

Walker Sanders
Third Vice Chairman
Enron Corporation
Houston, Texas

Patrick E. Clarke
Immediate Past Chairman
Washington Gas Light Company
Springfield, Virginia

John P. Erickson
Section Staff Executive
American Gas Association
Arlington, Virginia

A.G.A. STAFF

Michael Baly III
President

John P. Erickson
Staff Vice President
Operating and Engineering
Services

Jimmie A. Hicks
Director
Operating Section Services

Larry T. Ingels
Director
Utilization Engineering

Robert L. Parker
Manager, Engineering Series
Publications

Philip S. Runge
Manager, Utilization and
Engineering Services

Lori S. Traweek
Manager, Engineering Services

ENGLISH/METRIC CONVERSION FACTORS

An effort has been made in this book to provide International Standard (SI) metric equivalents in parentheses following English units. However, where the editor has, for whatever reason, not provided metric equivalents, the following table provides conversion factors for units in common use in the natural gas industry. Factors are carried to significant figures adequate to most practical applications. To avoid confusion, it is SI convention to replace commas (periods in some countries) by a half space every thousands place to the left or right of the decimal place.

To convert from	to	Multiply by
A		
acre (International Measure)	ha	0.404 685 6
acre (U.S. Survey Measure)	ha	0.404 687 3
atm	kPa	101.325*
atm	bar	1.013 25*
B		
bbl (oil)	m^3	0.158 987 3
Btu_{59}	kJ	1.054 804
Btu_{IT}	kJ	1.055 056
Btu_{IT}/bbl	MJ/m^3 (kJ/L)	0.006 636 102
Btu_{IT}/ft^3	MJ/m^3	0.037 258 95

---

* Conversion factors shown with an asterisk are exact.

Btu$_{IT}$/gal	MJ/m³	0.278 716 3
Btu$_{59}$/h	W	0.293 001 1
Btu$_{59}$/lb	kJ/kg	2.325 445
Btu$_{IT}$/lb	kJ/kg	2.326*
bushel, dry	L	35.239 07

C

cal$_{IT}$	J	4.186 8*
ft³ (cf)	dm³	2.831 685
ft³, dry	m³ (st)	0.022 328
ft³, sat	m³ (st)	0.027 835
ft³, sat (30 in. Hg)	m³ (st)	0.027 844
in.³	cm³	16.387 06
yd³	m³	0.764 554 9

D

dram, fluid	mL	3.696 691

F

ft (International Measure)	m	0.304 8*
ft (U.S. Survey Measure)	m	0.304 800 6

G

gal (U.S.)	L	3.785 412
gill	mL	118.294 1
gr	mg	64.798 91
gr/ft³	g/m³	228.835 2
gr/gal (U.S.)	g/m³	17.118 06

H

hp (electric)	kW	0.746*
hp (boiler)	MJ/h	35.314 2
hp (metric)	kW	0.735 499

I

in. (International Measure)	mm	0.025 4*
in. wc (60°)	kPa	0.248 843

M

mi (International Measure)	km	1.609 344*
mi (U.S. Survey Measure)	km	1.609 347

---

* Conversion factors shown with an asterisk are exact.

O

oz	g	28.349 53
oz, fluid	mL	29.573 53
oz/in.³	kg/m³	1 729.994

P

peck, dry	L	8.809 768
pt, dry	L	0.550 610 5
pt, liquid	L	0.473 176 5
lb	kg	0.453 592 4
lb/ft³	kg/m³	16.018 46
lb/in.² (psi)	kPa	6.894 757

Q

qt, dry	L	1.101 221
qt, liquid	L	0.946 352 9

R

rod (U.S. Survey Measure)	m	5.029 210

S

ft² (International Measure)	m²	0.092 903 04*
ft² (U.S. Survey Measure)	m²	0.092 903 41
in.² (International Measure)	mm²	645.16*
mi² (International Measure)	km²	2.589 988
yd²	m²	0.836 127 4
scf, sat (30 in. Hg)	m³ (st)	0.027 844

T

MBtu h 59°F	kW	293.001 1
MBtu h 59°F	MJ/h	1 054.804
thm (U.S.) 59°F	MJ	105.480 4
t, long	Mg	1.016 047
t/h (Mg/h)	t (short)/h	1.102 311
t, short	Mg	0.907 184 7
t, (short)/h	Mg/h (t/h)	0.907 184 7
T(°F)−32	T°C	0.555 555 6

---

* Conversion factors shown with an asterisk are exact.

Y

yd (International Measure)	m	0.914 4*
yd (U.S. Survey Measure)	m	0.914 401 8

---

* Conversion factors shown with an asterisk are exact.

Part 1.

UNDERGROUND STORAGE OF NATURAL GAS

Chapter 1

Introduction

This book in the Gas Engineering and Operating Practices Series is about underground storage of natural gas in its gaseous state. Gas is stored in many different types of underground reservoirs throughout the United States and Canada. Variations in reservoir type, location, and use result in widely differing practices and engineering procedures.

There are many handbooks and textbooks available for in-depth treatment of underground storage, reservoir engineering, and geology. No attempt is made in this book to duplicate these efforts.

This book provides general guidelines for the design, construction, operation, and maintenance of gas storage reservoirs, and more specifically gas storage wells. These guidelines were developed based on current technology and practice and should not be construed to reflect on facilities developed earlier. In many instances these guidelines represent a synthesis of many opinions and methods and are by no means the only available alternative; local conditions and practices must be considered along with the project objectives. It is expected that continued development of new technology will necessitate continued modification of these guidelines in the future.

PURPOSE OF UNDERGROUND STORAGE

Natural gas production is constant throughout the year, except for emergency shut-ins, which are usually due to the weather. Natural gas usage, on the other hand, is cyclic, with peak usage in the winter; consequently, storage is the only practical way to meet peak demands and avoid shortages for consumers. Gas is stored during low usage periods, helping gas pipeline companies maintain high levels of throughput efficiency. Then, in peak usage periods or during emergency production shut-ins caused by events like hurricanes and cold weather, gas is withdrawn from storage to supplement production.

HISTORY OF STORAGE
IN THE UNITED STATES

Natural gas was first stored underground successfully in 1915 in Canada, where a depleted gas reservoir was developed for peak requirements. An earlier attempt to dewater an abandoned salt mine for gas storage had failed.

In 1916 the first successful underground storage project in the United States was developed by Iroquois Gas Company (a National Fuel affiliate); that project, the depleted Zoar Field, is still in use today.

The first oil and gas field developed for storage of natural gas was the Fink Field in Lewis County, West Virginia, developed in 1941 by Hope Natural Gas Company (Consolidated Gas Supply Corporation). The first depleted oil reservoir for natural gas storage was started by Lone Star Gas Company, which developed the New York City Field in Clay County, Texas, for storage in 1954.

Louisville Gas and Electric Company started experimenting on the storage of natural gas in a water-bearing sand aquifer in 1931 and developed the Doe Run Field in 1946. This Meade County, Kentucky, project is the oldest aquifer storage field operating in the United States, although it no longer exhibits the water-drive characteristics of an aquifer.

Porous and permeable rock provided by nature was the only storage host of natural gas until 1961, when cavities in bedded salt were developed by Southeastern Michigan Gas Company in St. Clair County, Michigan. Leached cavern storage in salt dome intrusions was developed by Transcontinental Gas Pipe Line Corporation in 1970 in Covington County, Mississippi (domes occur principally in the Gulf Coast region of the United States).

In 1959 mined cavern storage of natural gas was introduced by the Public Service Company of Colorado, which converted an abandoned coal mine in Jefferson County, Colorado, into a storage reservoir.

No mine has yet been excavated solely for natural gas storage, although extensive studies have been conducted.

TYPES OF STORAGE

Seven types of underground storage are discussed in this book: aquifers, leached caverns, mined caverns, and depleted reservoirs,

MINED CAVERNS

Mined cavern storage in the free world today is done only in abandoned coal mines. Three abandoned coal mines are in use for storage today, one in the United States and two in Belgium. Mined cavern storage has the same high deliverability as leached cavern storage because it is also a closed container with infinite permeability. The storage operation is excellent for peak-shaving requirements because the cycle turnaround is so fast. The gas withdrawn during a high peak day can be replaced virtually overnight, depending on gas supply and operating conditions.

The amount of gas that can be stored depends on the size of the caverns and the pressure maintained in them. A mined cavern is maintained at a low operating pressure; reinjection is therefore accomplished by free flow, because the pipeline pressures are higher than the cavern pressures. Withdrawal is done by compression for the same reason. Care must be exercised not to exceed the hydrostatic limits of the host rock, because current mined storage is very shallow compared to other storage methods, and the operating pressure of the mine must be kept at low levels.

The conventionally mined cavern is one of the most expensive methods of storage today, partly because of several factors that are not controlled by the gas companies:

- When ore is being removed from a mine, no care is taken to prevent cracks, microfissures, or other openings that could contribute later to the loss of storage gas.
- There are natural fissures or openings in the rock that are characteristically associated with coal deposits.
- The shafts that are created in coal mines for ingress and egress must also be sealed to ensure against loss.
- The pillars that support the roof of the cavern can cause problems with well deliverability. The injection-withdrawal wells must be drilled into the cavern, much like any producing well is drilled in an active oil or gas field. Deliverability is severely restricted and may even be eliminated if a well is drilled into a pillar, because the pillar diameter is so large that a perforating gun cannot shoot completely through it; as a result, only the gas that can seep through the microcracks in the rock can reach the wellbore.

That only one mined cavern storage facility operates in the United States is an indication of how few favorable areas exist. Although abandoned coal mines are plentiful, they are not always convertible to gas storage. No mined cavern in existence was mined solely for storage (leached caverns excluded). Economics, need, and host rock dictate the favorable areas for mined cavern storage; a favorable area has base-ment rock close enough to the surface to be mined for storage. Examples are the granites of Georgia and the schists of New Jersey.

DEPLETED RESERVOIRS

Depleted reservoirs are the most common of all storage reservoirs and probably the least expensive to develop. The types of depleted reservoirs discussed in this book (condensate reservoirs, high- and low-permeability gas reservoirs, and high-permeability oil reservoirs) all have the same characteristics—with a few exceptions, which are discussed in detail in the case studies.

The rock types found in depleted reservoirs are clastic and nonclastic. The nonclastic rocks are chemical precipitates and the clastic rocks are depositional, such as sandstone, conglomerate, shale. (See also the "Geology" section.) The reservoir traps (i.e., caprocks and faults) confine oil and gas. When the reservoir is depleted and converted to storage, the traps contain the storage gas.

Storage gas may migrate if the reservoir storage pressure exceeds the original field pressure. A producing field's original pressure may or may not be at the maximum threshold. Because of natural compression caused by faulting or other geologic movement, some fields contain only the amount of fluids that can be trapped at the discovery pressure; the fluids present at higher pressures have already migrated to other host rocks. Some fields can be delta pressured because they have never reached the maximum pressure possible during geologic history. The storage capacity of a field that can be delta pressured is greater than the original production limits, and pressure-volume-temperature (PVT) equations (see the "Reservoir Engineering" section) can be used to predict the maximum volume of storage space.

Production and storage wells are another source of gas leaks from the depleted reservoir. Old production wells must be plugged and abandoned in a manner that prevents gas from migrating into shallower sands. New storage wells must be completed with good cement integrity, which prevents migration behind the casing.

Production records are used to determine the size of the field, and geological and geophysical records help determine the actual limits of the reservoir. The surface boundaries must be established in such a manner that no company can accidentally or intentionally drill into the storage stratum without approval of the storage company.

Many favorable areas exist in the United States for storage of gas in depleted reservoirs. Storage potential exists anywhere there is an active or depleted production field.

Chapter 2

DESIGN, CONSTRUCTION, OPERATION, AND MAINTENANCE CONSIDERATIONS

PRELIMINARY PLANNING

Before starting a storage development program, the operator should:
- Collect and analyze all available engineering and geological data as a basis for design criteria. Sources of such data are drilling companies, field operators, state and federal records, scouting records, and people living in the area. In the case of a storage reservoir such as an aquifer, methods such as a pilot program of drilling wells or a geophysical program, are necessary.
- Protect all freshwater zones with cemented casing to prevent gas from migrating into the freshwater zones and to prevent the commingling of freshwater zones, as required by state laws.
- Be prepared to prevent hydrocarbons from migrating between zones and storage wells and, in addition, be prepared to control any unusual pressure conditions.
- Prepare for possible lost circulation by having sufficient material and equipment. Such preparation should be adequate to preclude loss of control of any upper gas zones.
- Prepare adequately to minimize the hazards from sloughing, unconsolidated, and creep zones.
- Make adequate preparation to control any operation that would be an unreasonable detriment to the overall ecological environment.

- Establish and record the number and condition of the dry holes and plugged and abandoned wells in the reservoir and in the protective buffer area, including water wells. If any conditions of the dry holes or plugged and abandoned wells could affect the integrity of the reservoir, those wells should be reentered and properly reconditioned or plugged and abandoned, including those in the protective buffer area.
- Establish a minimal buffer zone as a protective area surrounding the storage reservoir areal extent. The storage reservoir areal extent should be established by an applicable technique such as dry holes, geophysics, or reservoir analysis.
- Determine the maximum reservoir pressure by means of the reservoir geometry, overburden density, and the reservoir stratigraphy.
- Determine the pressure-volume relationship by means of a recognized engineering method. This pressure-volume relationship should be monitored from season to season to confirm the integrity of the storage reservoir and related facilities.
- Review local, state, and federal regulations.

DETERMINING STORAGE NEEDS

The design criteria for a storage reservoir must be based on consumer demands, availability of storage fields, and other economic considerations. Storage reservoirs can be used for pipeline flexibility, peak shaving, emergency weather shut-ins of production fields, storage of low-cost supply during low-demand periods, seasonal demands, and so forth. The design should be flexible and allow for future changes due to new and different demands.

Deliverability

The ability to deliver gas depends on reservoir properties. The planning staff must consider the facilities available and the deliverability needed in order to select the type of storage that will best satisfy the delivery criteria.

Volume

A field that is needed for large volume storage may be located a great distance from delivery points. These fields generally have good but not high deliverability and are best suited for storage for pipeline flexibility and seasonal demands.

Alternatives to Storage

When a storage field is not economically available, some alternatives must be considered. Liquefied natural gas storage is one alternative. Line packing, another form of storage, is useful but only in comparatively small quantities. Consumer conservation may be encouraged but usually cannot be enforced, especially during extremely cold weather. Consumer cutbacks may become necessary, but this is an extreme measure and should be used only in case of emergencies.

SITE SELECTION

A storage field cannot always be situated in an ideal location. Some fields, depending on their use, may not need to be close to consumer delivery points, whereas others should be within a few miles. Future development of the field greatly depends on the surrounding population. Because future population growth must be considered when a storage facility is being designed, site selection is extremely important.

Proximity to Market

Fields used strictly for volume storage do not need to be close to the sales point. Storage that is used for peak shaving should be close to the market area to deliver the large amount of gas needed during peak periods. Gas delivery from a pipeline is a function of pressure, pipe area, and length. Compressor stations are used along the pipeline to enhance throughput.

Proximity to Pipeline

It is desirable, because of cost and deliverability, to select a storage facility that is close to a pipeline. The cost of laying a lengthy pipeline to tie into an existing transmission line is very high. Delivery, as previously discussed, is affected by the length of the line to the destination. A storage field that is too far from the transmission main line cannot deliver gas at a reasonable cost. If the criterion of economical gas delivery is not met, an alternative must be found.

Availability of Suitable Formations

Production methods often preclude a field from storage because secondary and tertiary production procedures make conversion economically impossible. Some companies use advance planning during production practices and design storage fields before depletion, making conversion more economical. A storage reservoir formation may be found naturally, such as depleted production fields, or designed and constructed, such as mined or leached cavern storage. The latter formations have infinite permeability and porosity with no water influx, but not all storage formations have these unique characteristics. Many successful production fields are not economically convertible to storage because of their natural geological formation.

GEOLOGY

The study of the nature of the earth's crust combined with a study of its ability to store petroleum is extremely important to the storage engineer. This knowledge of petroleum geology is used to find and produce undiscovered oil and gas and eventually convert a depleted field to storage. The engineering staff of a storage department must go even further to develop different types of storage fields such as leached cavern storage, mined cavern storage, and aquifer storage.

The three basic rock types are extremely important in the design of any storage field. The rock types are clastic, nonclastic or precipitates, and miscellaneous. The clastic rocks are sedimentary, consisting of fragments of other rocks that have eroded and settled together. The nonclastic rocks are precipitates of chemical reactions consisting mainly of limestones and dolomites. Clastic and nonclastic rocks are the basic components of almost all depleted reservoirs and aquifers. The third rock type is miscellaneous, commonly referred to as basement rocks. Basement rocks almost never contain oil and gas that can be produced in large enough amounts to be economical. Some rare cases do exist, but only because the fluids migrated from the original deposit into cracks and fissures in the basement rock. Basement rocks include the igneous and metamorphic rocks and their mixtures. The basement rocks are ideal for mined cavern storage because they are impermeable and impervious.

LAND AND RIGHT-OF-WAY

Where operating conditions dictate considerable use of the surface, the extent of surface ownership should be considered. In the case of depleted production fields, where some production may be realized during cycling operations, mineral rights must also be negotiated. The base gas in a storage field is sometimes made up of nonrecoverable gas left in the reservoir after producing operations have ceased. This gas, although unrecoverable, becomes part of the storage operation and therefore is usually assessed a value. A properly negotiated land acquisition encompasses all these contingencies. Access to the storage field for delivery and withdrawal purposes must be developed in close association with the field acquisition.

REGULATORY REQUIREMENTS

Storage fields, when used for interstate transmission, are regulated by the Federal Energy Regulatory Commission (FERC). State agencies may have some jurisdiction over different aspects of the operation of storage fields and have even more regulatory powers over storage facilities for intrastate transmission systems. Filing procedures have been established by these agencies, and early contact should be made with them for full details.

RESERVOIR ENGINEERING

The storage engineer is concerned with developing, maintaining, and cycling a storage field as economically as possible. Field development starts when the first production well is drilled and continues until the field is abandoned for storage as well as production. Development and maintenance of a storage field are almost interchangeable from an engineering standpoint. Well abandonment, completion, recompletion, and workovers are a continuous operation and all are critical engineering practices. The size of a depleted field reservoir, although well established from production history, must be continually monitored to verify inventory. Reservoir size in aquifers must be "designed" because of the movement of the water face; no previous production exists that can be used to establish the limits. In the case of mined cavern storage, cavern sizes are set, but salt caverns are plastic and must be continuously evaluated. All aspects of storage are different, but all use the same PVT relationships to establish and monitor inventory.

The general equation for calculating pore volume in a storage reservoir is

$$V_1 = \frac{V_2}{\frac{T_b}{P_b}\left(\frac{P_1}{T_1 Z_1} - \frac{P_2}{T_2 Z_2}\right)} \qquad\qquad \text{Eq. 1}$$

where:

V_1 = pore volume, cf (m³)
V_2 = volume of gas injected or withdrawn for test period, cf(m³)
P_b = base pressure, psia (kPa)
T_b = base temperature, °R (K)
P_1 = initial formation pressure, psia (kPa)
T_1 = initial formation temperature, °R (K)
Z_1 = compressibility factor, initial conditions
P_2 = final formation pressure, psia (kPa)
T_2 = final formation temperature, °R (K)
Z_2 = compressibility factor, final conditions

Then, gas in place equals G_p as follows:

$$G_p = \frac{T_b}{P_b} \left(\frac{P_2}{T_2 Z_2} \right) \left(\frac{V_1}{1 \times 10^9} \right), \text{ Billion Standard Cubic Feet (m³) at standard conditions} \qquad \text{Eq. 2}$$

These formulas are valid for volumetric reservoirs with no water influx; many depleted fields and leached cavern storage projects are good examples. Because aquifers and many other depleted reservoirs do have water influx, however, a term must be added to the formula. The pore volume with water influx may be calculated from the following formula:

$$V_2 = \frac{T_b}{P_b} \left[\frac{P_1 V_1}{Z_1 T_1} - \frac{P_2 (V_1 - W_e + B_w W_p)}{Z_2 T_2} \right], \text{ cf (m³)} \qquad \text{Eq. 3}$$

where:
W_e = water entering the reservoir, cf(m³)
W_p = water produced from reservoir, cf(m³)
B_w = formation volume factor for water, bbl

Eq. 1 was initially developed for a producing reservoir, where the initial pressure is always greater than the final pressure. In a storage reservoir, where the inventory changes cyclically, the initial and final pressure difference vary positively or negatively, depending on when the pressure surveys are run—after injection or after withdrawal. Because no negative pore volume can exist, an absolute value is assumed in the calculation. The formulas can be used for British units or SI units, but the units must be constant throughout.

RESERVOIR SYSTEM

Hydrocarbons migrate and accumulate in oil and gas reservoirs because of the presence of traps. When depleted of their oil and or gas, these traps may be used for underground storage purposes. Folding (Fig. 1) and faulting (Fig. 2) of subsurface rocks create deformities where oil or gas can accumulate in structural traps. Stratigraphic traps (Fig. 3) are formed by lithologic changes due to changes of depositional environment or later dissolution (diagenesis) of the hydrocarbon bearing rocks. Many traps are combinations of structural and stratigraphic features. Traps can be a variety of sizes with an equal variety of porosity and permeability. Aquifer storage development requires similar geologic features. Mined and leached caverns are somewhat different in makeup. The traps in these instances are manmade and are not included in the natural stratigraphy.

Much importance must be placed on the volume of oil or gas in place in a reservoir. When no previous production records are available, pore space must be calculated. The equation for volumetric oil in place is

$$N = \frac{7\,758\,\phi\,(1 - S_w)}{B_o}, \text{ bbl/acre-ft} \qquad\qquad \text{Eq. 4}$$

or

$$N = \frac{10\,000\,\phi\,(1 - S_w)}{B_o}, \text{ m}^3\text{/ha-m}$$

Courtesy of the California Department of Conservation, Division of Oil and Gas

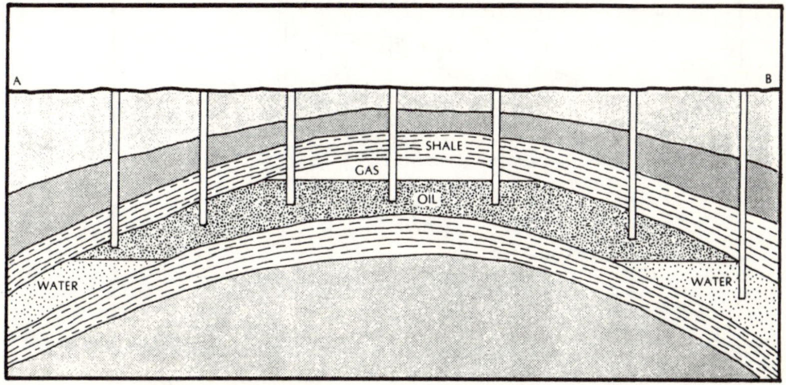

Figure 1. **Longitudinal view of a typical folded structural trap, in this case, an anticline.**

Courtesy of the California Department of Conservation, Division of Oil and Gas

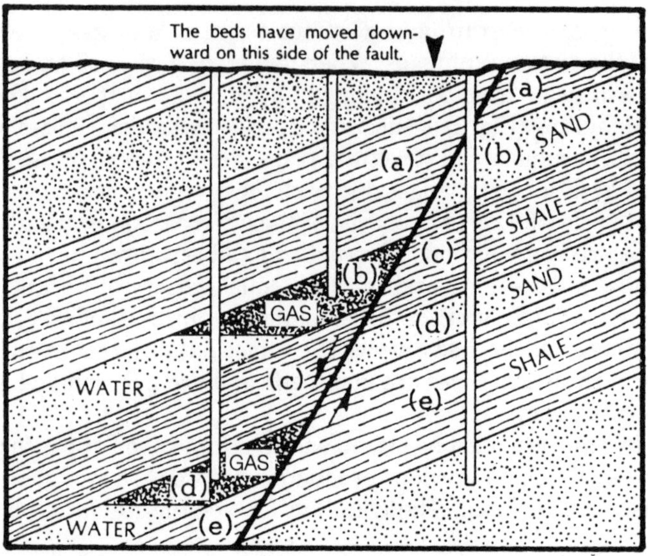

Figure 2. Example of gas entrapped by faulting. Note the impermeable beds now opposite the gas zones.

Courtesy of the California Department of Conservation, Division of Oil and Gas

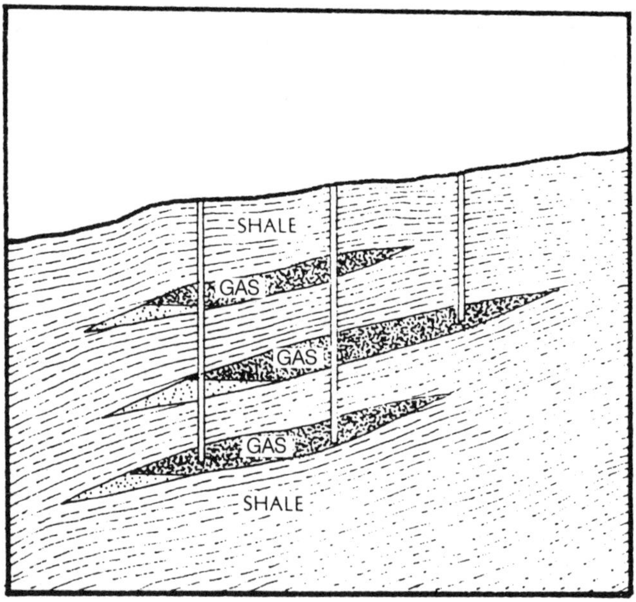

Figure 3. Stratigraphic trap with sand lenses interspersed in a shale bed. The shale acts as a permeability barrier.

where:

N = tank oil in place, as defined in Eq. 4

S_w = water saturation

B_o = formation volume factor, $\dfrac{\text{reservoir bbl}}{\text{tank bbl}}$

ϕ = total porosity

The equation is slightly different for a gas reservoir, because the gas in place must be measured in cubic feet. This equation is

$$G = 43\,560\,\phi\,(1 - S_w) \times \frac{T_b}{P_b}\,\frac{P_i}{Z_i T_i}\,, \quad \text{Mcf/acre-ft} \qquad \text{Eq. 5}$$

or

$$G = 10\,000\,\phi\,(1 - S_w) \times \frac{T_b}{P_b}\,\frac{P_i}{Z_i T_i}\,, \quad \text{m}^3/\text{ha-m}$$

where:

G = gas in place at standard conditions as defined in Eq. 5

ϕ = porosity, percent

T_b = temperature at standard conditions, °R (K)

P_b = pressure at standard conditions, psia (kPa)

P_i = initial reservoir pressure, psia (kPa)

T_i = initial reservoir temperature, °R (K)

Z_i = initial compressibility factor, gas

S_w = water saturation

and,

43 560 = Number of cf/acre-ft

(10 000 = Number of m³/ha-m)

Porosity is calculated by measurement of core samples as follows:

$$\phi = \frac{V_p}{V_b} \qquad \text{Eq. 6}$$

where:

ϕ = porosity

V_p = pore volume of core sample, cm³

V_b = bulk volume of core sample, cm³

$$V_p = \frac{W_{sat} - W_{dry}}{g} \qquad \text{Eq. 7}$$

where:

W_{sat} = weight of the sample saturated with brine, gm
W_{dry} = weight of a clean dry sample, gm
   g = specific gravity of brine, gm/cm³

The pore volume is the factor of most interest here. These equations estimate the amount of fluid the reservoir will hold. The ability of the fluid to flow through the rock is called permeability; a reservoir can have high porosity, hence a tremendous ability to store fluid; but if it does not possess permeability, it is useless as a production or storage reservoir. A good example is a lava flow. There is a large bubble structure in lava but little interconnection and therefore almost no permeability. A core sample that "bleeds" oil when pulled from the core barrel is a good indication of a poor well. The rock sample obviously is porous, but the permeability is so low that the drilling fluid did not wash the oil from the core as it was pulled out of the wellbore. New fracturing techniques may help a production well in this case, but the expense is high. A field that is depleted under these conditions would probably not be a good storage reservoir.

GAS INVENTORY

Gas inventory consists of two categories of storage gas, base gas and top gas. Each classification has significant importance, one for pressure maintenance and the other for sales.

Base gas may be either native or injected. A "base" of gas must exist to maintain a pressure base and ensure deliverability during peak drawdown periods. If a depleted field is below this pressure before development for storage, additional gas must be injected to raise the pressure base. This gas is not intended for sale during regular cycling periods and becomes part of the base gas system. Frequently a reservoir is converted to storage before the native gas is depleted to a point below a working pressure level and no gas needs to be injected; all the base gas is native. However, aquifer and mined and leached cavern storage have no native gas associated with them, and all base gas is injected. All of this gas can be recovered by injecting water into the cavern and producing the gas. Some methane does exist in coal mines that have been converted to storage, but (except for one mine in Belgium) not in quantities large enough to create a pressure base. A mine in Belgium, which no longer produces coal, produces so much methane that the owners have not yet converted it to storage.

Definition

Different storage fields and concepts rarely lend themselves to identical classifications. The four major storage field types are defined by different methods. Mined and leached cavern storage, although they are closed container-type reservoirs, do present some problems with definition. Leached salt caverns have a tendency to shrink because of the plastic flow of salt.

Depleted reservoirs have numerous characteristics, among them varying porosity, permeability, and formation thickness; each contributes to a degree of difficulty concerning reservoir definition. Aquifer definition is probably the most difficult of all. Porosity, permeability, formation thickness, and a constantly retreating or encroaching water face make an aquifer storage reservoir extremely hard to define.

Inventory Monitoring

For a high degree of confidence to be maintained in a storage field, its integrity must be evaluated periodically, gas inventory must be monitored, and migration must be controlled. Inventory control may consist of determining if any migration exists or, as in the case of some aquifers, doing remedial work to find and produce gas that has been lost from the storage field.

Determination

Before a remedial program is started, a determination is made that gas has migrated from the reservoir. This determination can be made by simple pressure monitoring, observing gas leakage at the surface, or using more detailed and intricate reservoir engineering techniques. It is a good idea to maintain observation wells in shallow sands above the reservoir. Any escaped gas migrates to areas of low pressure, and unless it encounters a trap, it eventually seeps at the surface. Pressure monitoring of an observation well usually indicates a pressure change before the operator realizes the loss. Using pressure gauges and periodic logging practices helps prevent a problem that could endanger the storage reservoir as well as the safety of personnel and equipment on the surface.

Verification and Quantification of Losses

Because of the interest of various government agencies such as FERC, the Environmental Protection Agency (EPA), and related state agencies, and also because of the economics involved when a gas loss is experienced, losses must be quantified as well as verified. Most reservoir engineering formulas are for quantification of gas inventories. Other means of quantifying, such as accurate gas measurement, are used to establish loss, assess value, and determine remedial recovery processes. Migrated gas may be recovered unless the recovery processes are not economical. The migrated gas must then be written off and the inventory must be adjusted accordingly. In some cases of unaccounted losses, gas migration cannot be proven; examples are surface valve leaks, piping leaks, casing leaks to the surface, and poor measurement techniques. (Poor measurement can result in an indicated gas overage as well, which is also undesirable.)

DELIVERABILITY

"Absolute open flow" (AOF) is the term used to define a gas well's ability to flow if the only pressure against the sandface is atmospheric. The formation and sandface pressures are valid terms only when used in relation to depleted fields and aquifers. As previously stated, mined and leached cavern storage reservoirs are simply closed containers, and a general flow equation does not define their abilities. A deliverability equation is discussed in detail in the leached cavern case study (see Part 2, Chapter 2).

Well Deliverability

Individual well deliverabilities are calculated by plotting on a log-log graph the difference between the squares of the sandface pressure and the formation pressure versus different flow rates. The point of intersection of the line passing through or near a number of points and the square of the original reservoir pressure is said to be the AOF potential for a particular well.

The formula for determining the static reservoir pressure (P_f) of a reservoir is as follows:

$$P_f = (P_w)\,(e^s),\ \text{psia (kPa)} \qquad\qquad \text{Eq. 8}$$

where:

P_f = shut-in reservoir pressure, psia (kPa)
P_w = shut-in wellhead pressure, psia (kPa)

and

$$s = \frac{0.018\ 77\ (G)(H)}{T_a Z_a}$$

where:

G = specific gravity of the gas column

H = vertical distance from reservoir to surface, feet

T_a = average temperature of the gas column, °R

Z_a = average compressibility factor of the gas column

To convert from psia to kPa, multiply psia by 6.894.

This very accurate formula is used by several agencies for calculating formation pressures. The formula involves trial and error, because the formation pressure must be initially estimated and an average Z-factor calculated and entered into the formula. The Z-factor is interated and changed each time until the reservoir pressure is eventually estimated correctly.

The following formation pressure, P_s, is a bit more difficult to calculate. The following formula is used, and iterations must also be made because of the changing Z-factor. This equation has the same root as Eq. 8, with a friction pressure loss term added.

$$P_s^2 = P_t^2 e^{2s} + \frac{0.002\ 68}{D^5}\ f\ (T_a Z_a Q)^2 \frac{L}{H}\ (e^{2s-1}),\ \text{psia} \qquad \text{Eq. 9}$$

where

P_s = flowing reservoir pressure, psia

P_t = flowing wellhead pressure, psia

D = flow string diameter, inches

f = Moody friction factor at average flowstring pressure and temperature

Q = gas flow rate, MMcfd at 14.65 psia and 60°F

L = length of flow string, feet

and other variables are the same as in Eq. 8. Note that e is taken to the power of 2s, or twice the value of s used in Eq. 8.

This formula can be used to estimate the sandface pressure on injection as well as withdrawal by changing the negative sign before 15.1 (176.02) to a positive sign. Well deliverabilities can then be calculated on the injection mode as well as the withdrawal mode.

The difference between the squares of the formation pressure and the sandface pressure must then be plotted versus the flow for each pressure. See the graph in Figure 4 for more detail. The point of intersection with the square of the original reservoir pressure is the AOF potential.

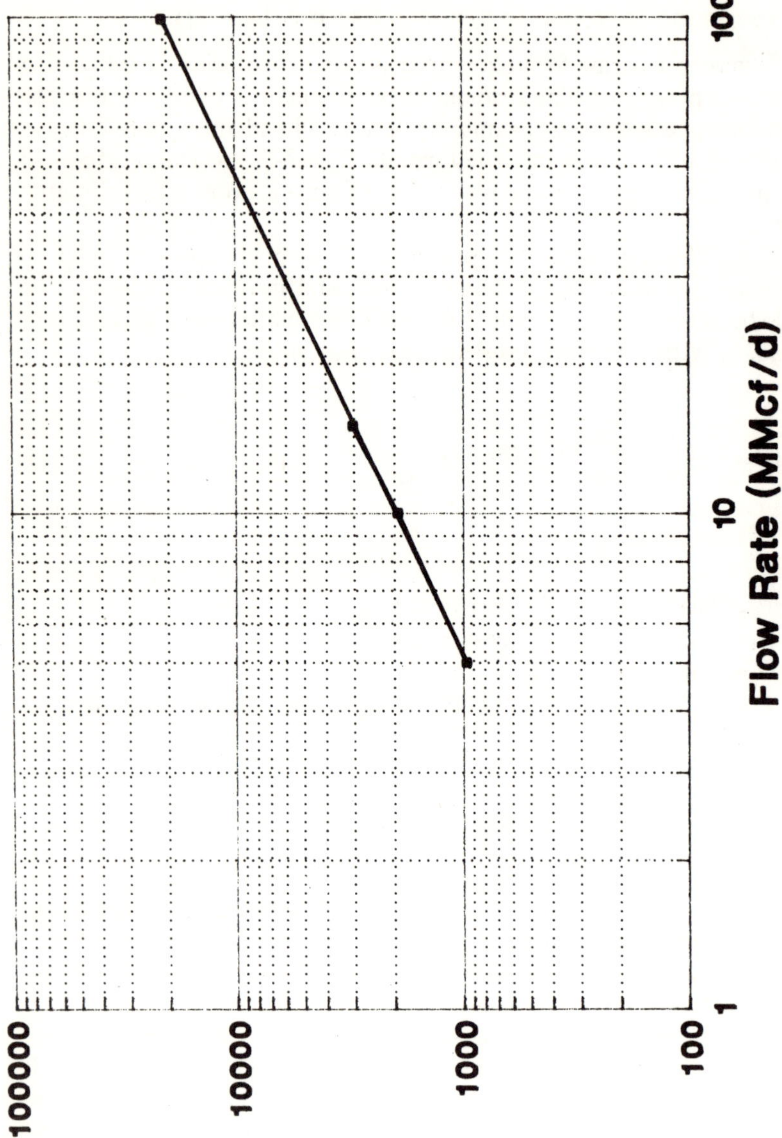

Figure 4. Sample log-log graph of a deliverability test. AOF=99 MMcf/d.

The slope of the line N through the points should fall between a slope of N=1 and a slope of N=0.5. Some operating conditions cause a deviation from this slope, but this phenomenon is not discussed here. The slope of the line is determined by the flowing conditions, turbulent and laminar. A well that flows in a totally turbulent condition has a slope of 0.5, whereas a well with laminar steady-state flow has a slope of 1.0. If the slope of the line is outside these parameters, it can be adjusted to calculate the AOF. See Figures 5 and 6 for example adjustments.

Field Deliverability

The total field potential deliverability is the sum of all the individual well deliverabilities in a field. Some companies prefer to express this potential as an average field deliverability by using either a simple arithmetic average or a weighted average.

FACILITIES

SUBSURFACE FACILITIES

The process of the delivery of gas to a pipeline starts long before the first hole is drilled and the reservoir formation is penetrated. Preliminary planning necessary for a successful well completion incorporates well design, drilling, down-hole equipment, and completing practices to ensure a minimum of errors.

Well Design

All wellhead equipment should conform to current American Petroleum Institute (API) Standard 6-A (see Appendix A).

Wellhead equipment should be designed to accommodate the maximum gas pressure and stimulation treating pressure to which the equipment may be subjected.

The wellhead should be equipped with enough valves to isolate the well from the pipelines. Consideration should be given to installing the necessary connections and equipment on the wellhead for emergency control, including a master valve that is protected on both sides to prevent failure.

Wellhead equipment should be designed so that full bore down-hole equipment can pass through it.

Depending on the degree of exposure to accidental hazards, the wellhead should be protected by a fence, guard rails, or other protective devices.

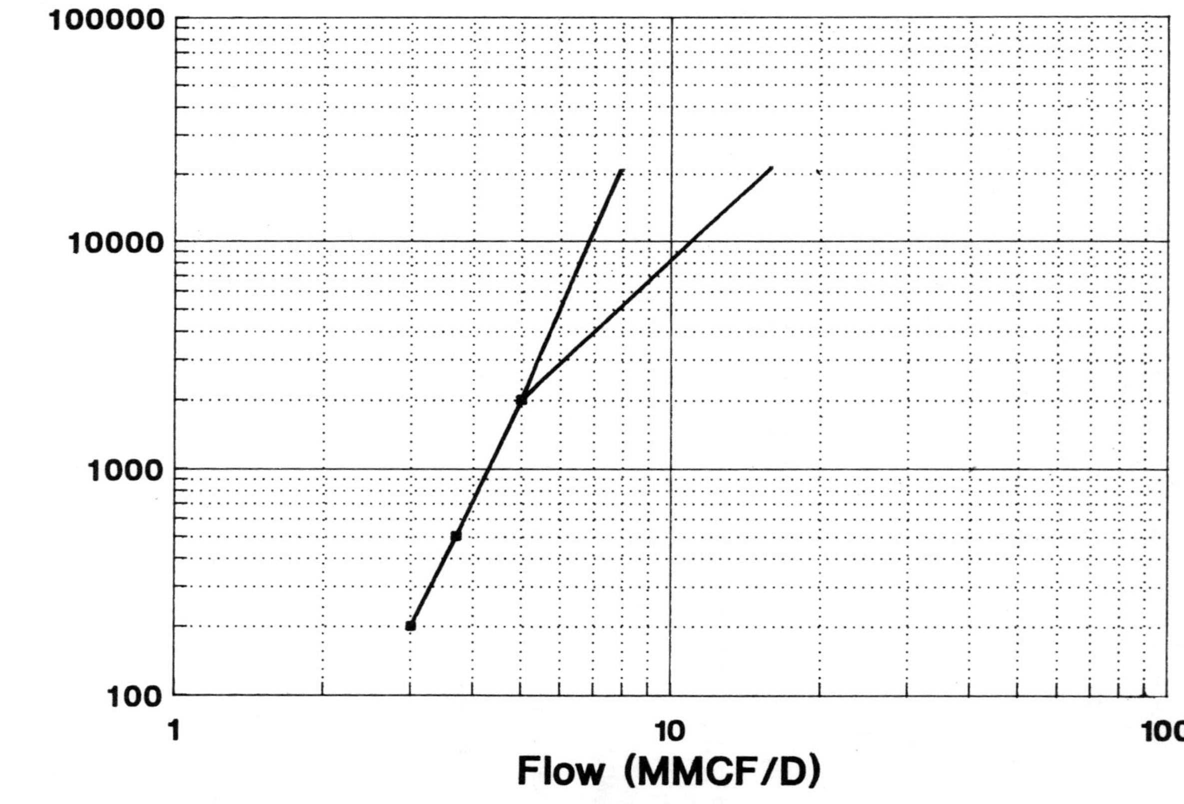

Figure 5. Sample log-log graph of a deliverability test. AOF=16.1 MMcF/d.

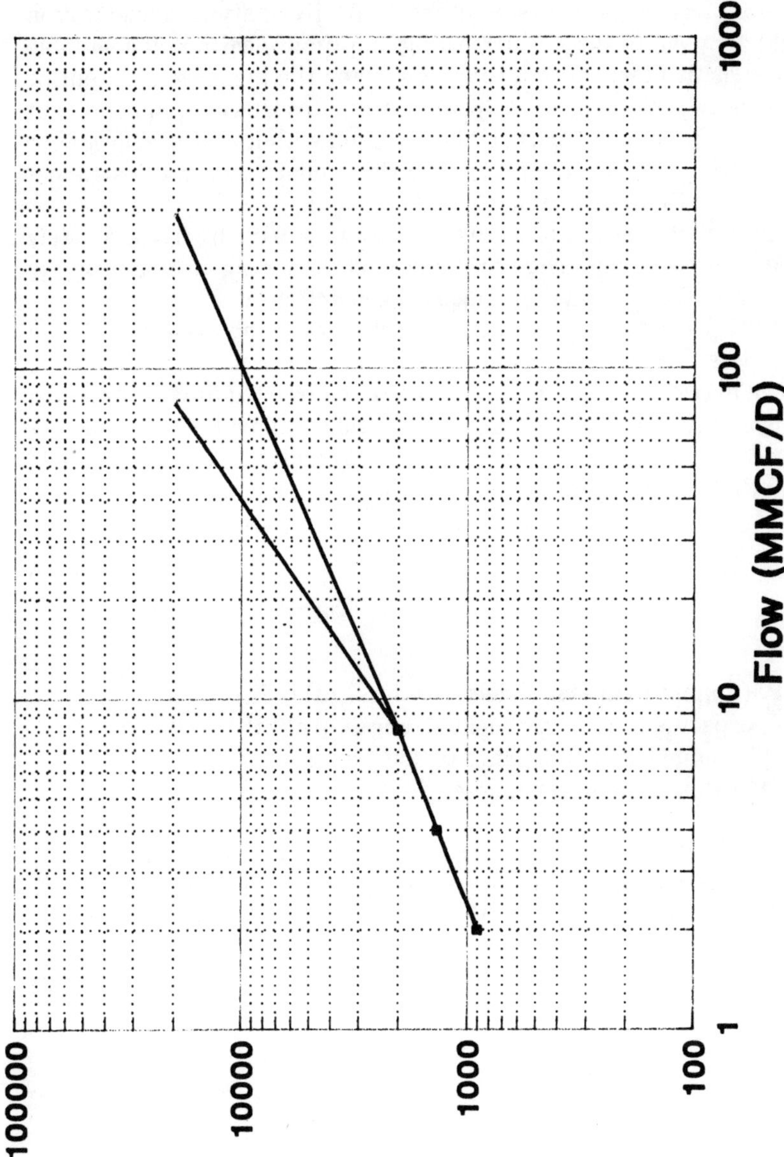

Figure 6. Sample log-log graph of a deliverability test. AOF= 85 MMcf/d.

In the design of wellhead equipment, consideration should be given to the use of ring joint flanges, raised-face flanges, or threaded wellhead equipment. Operating pressures and open flow may influence this design.

All casing to be used for storage well completions should be manufactured and inspected in accordance with the current API Specifications #5A, #5AC, and/or #5AX (see Appendix A). In addition, consideration should be given to further inspecting all primary casing by a suitable nondestructive test procedure to determine the presence of latent or other undesirable defects. Any additional subsurface equipment that becomes a permanent part of the injection-withdrawal casing and/or tubing string such as staging tools or float collars should at least meet applicable API specifications.

The casing should be sized so that one string of casing fits adequately inside another string. The reasons are so the casing can be properly centered and to provide an adequate cement sheath.

If the second string is to be cemented, then the clearance should not be less than approximately ½ inch.

The maximum gas pressure that is to be exerted in the completed casing string of a new storage well typically will be limited so as not to produce a stress in excess of 72 percent of specified minimum yield strength. This is expressed by the following formula:

$$P_i = .72 \, \frac{2Y_p t}{D} \qquad \text{Eq. 10}$$

where:

P_i = minimum internal yield pressure, psi (kPa)
Y_p = specified minimum yield strength, psi (kPa)
 t = nominal wall thickness, in. (cm)
D = nominal outside diameter, in. (cm)

The pipe body yield strength of new storage well casing typically will be approximately twice the deadweight (in air) of the casing. The following formula expresses the pipe body yield strength:

$$P_y = .7854 \, (D^2 - d^2) \, Y_p, \; \text{lb} \qquad \text{Eq. 11}$$
$$\text{or}$$
$$P_y = 8.009 \times 10^{-3} \, (D^2 - d^2) \, Y_p, \; \text{Kg}$$

where:

P_y = pipe body yield strength, lb (kg)
Y_p = specified minimum yield strength for pipe, psi (kPa)

D = specified outside diameter, in. (cm)
d = specified inside diameter, in. (cm)
P_y should be at least twice the weight of the casing.

The design unit stress of new storage well casing under tension at the minimum cross-sectional area of the casing typically will be limited so as not to exceed approximately 80 percent of the specified minimum yield strength. The following formula expresses the joint strength:

$$P_j = .95 A_{jp} \, UP, \text{ lb} \qquad\qquad \text{Eq. 12}$$

or

$$P_j = .32 A_{jp} \, UF, \text{ kg}$$

where:

P_j = minimum joint strength, lb (kg)
A_{jp} = cross-sectional area under the last perfect thread, $n^2(cm^2)$
UP = minimum ultimate strength, lb (kg)
P_j should not be greater than 80 percent of the minimum yield strength.

Collapse pressure will be reviewed and will typically be limited to 80 percent of the minimum resistance for the grade and weight of the new storage well casing used. The following formula expresses the yield strength collapse:

$$P_{yp} = 2 \, YP \left(\frac{D/t - 1}{(D/t)^2} \right) \qquad\qquad \text{Eq. 13}$$

where:
P = minimum yield strength collapse pressure, psi (kPa)
YP = minimum yield strength of pipe, psi (kPa)
D/t = intersection between yield strength collapse and plastic collapse, in./in. (cm/cm)

Before running casing, the operator should obtain adequate hole conditions (i.e., volume, content, etc.) to assure successful cement circulating conditions, reduced mud cake, and adequate annular fillup and bonding. Consideration should be given to circulating a volume of cement equal to approximately twice the hole displacement.

All casing to be used in storage well completions should be handled as set forth in the API Specification #5C1 (see Appendix A). In addition, the following should be considered:

- Thread lubricant specifications, as found in API Recommendations #5A2 (see Appendix A), should be considered. Consideration should be given to the manufacturer's specifications regarding special thread lubricant for their tubular products.
- To maintain the integrity of the casing string and ensure that the bottom joint does not back off, the operator should investigate using a thread locking compound on the casing shoe and every threaded connection to a point of the first threaded connection above the top of the cement to be left inside the casing at the end of the cementing procedure.
- Consideration should be given to using hydraulic power tongs in running the production string. Dogs used with the hydraulic power tongs should be selected so as to cause the least amount of damage to the casing-tubing and help to retard stress failure.
- The operator, using hydraulic power tongs, should ensure that the casing-tubing string is made up to the recommended torque as specified by the manufacturer and/or by API Recommended Practice #5C1 (see Appendix A). It should be noted that certain threads require a position makeup rather than a recommended torque.

Drilling

The following sequence of events is for onshore cased-hole completions. The intent here is not to give full details on drilling procedures, but rather a general discussion. It is advisable to refer to the many books about drilling for an in-depth detailed study on the subject. Many high-deliverability wells are open-hole completions with production casing set in the same manner. The open uncased storage interval can be undereamed and gravel packed to aid in deliverability and sand control.

A drilling permit is issued by state or federal agencies before a drilling program is started. Permits usually set time limitations, tract locations, depth, and principal target formations for the expected completion.

Site preparation must then be completed, including
- Leveling, shelling or graveling
- Laying a two- or three-layer oil field board pad
- Drilling a water well or making arrangements for a water supply
- Digging mud pits or moving in portable units
- Laying a fuel gas line if needed

The rig may now be rigged up and a rathole dug for the kelly and a mousehole dug for connections.

If desired, conductor pipe may be run by driving it with a hydraulic hammer, or a conventional hole may be drilled and the pipe cemented to the surface. When the conductor pipe is set, surface casing will be set according to government policies for maximum and minimum depth. The surface casing protects the freshwater strata from contamination during and after drilling. The surface casing is cemented to the surface by means of approved methods to ensure that the cement surrounds the pipe concentrically. The casing hanger flange (starting bowl) is usually installed while the cement is setting, either by screwing onto the casing or cutting the casing and welding the flange on top.

After the casing hanger flange is attached, the blowout preventer (BOP) is installed and tested, if required.

The casing is tested next by drilling the cement plug down to the cement shoe and building up pressure with mud, rig pumps, and BOP. The drilling continues, either to an intermediate casing set point or to total depth. The bits used are designed for longevity and rate of penetration. The bit size is selected to allow about a 1-in. annulus between the casing and formation. Bit cones and shapes of teeth and materials used vary so much that it is impossible to list them here.

The mud weight and properties are monitored and evaluated. Properties such as weight, viscosity, water loss, and gel strength are evaluated to prevent problems such as blowouts, lost circulation, and stuck pipe. The mud properties are checked for oil or gas shows, and when total depth is finally reached, the mud is thoroughly conditioned and the drill pipe is tripped out of the hole.

Logs are run at total depth to determine the target formation's top and base, porosity, connate water, and permeability. Sidewall cores can be taken after the hole is drilled, or conventional cores (which may give better results) can be taken. Conventional coring is more expensive because it consumes more rig time.

The drill pipe is laid down and loaded out after a final trip is made into the hole to condition the mud and tag bottom. The production casing can now be set. When the production casing string is cemented, a two-stage program may be used to prevent the hydrostatic head from fracturing the formation.

While the cement is hardening, the casing may be hung off and a work string picked up. After sufficient time has elapsed for the cement to harden, the cement plug inside the production string is drilled to within approximately 25 ft (7.62 m) of the casing shoe. After the work string is tripped out and the mud is conditioned, a cement bond log can be run to evaluate the quality and quantity of cement bonding.

When an acceptable isolation has been achieved, a packer is typically run; care must be used to prevent it from being set in a casing collar. Some packers are set by electric wireline, a service offered by most logging and perforating companies. Some packers are run in the hole on tubing and set by turning the tubing and setting down or lifting up while using the mud pump pressure. The packer should now be pressure tested for leakage.

After the packer test is accepted and the mud is displaced with a packer fluid, the work string is laid down and loaded out. The BOP can now be removed and the portion of the wellhead assembly that accepts the tubing hanger can be installed. The tubing with an appropriate length of packer seals is run and spaced out. Before the tubing is run, it should be checked for transverse and longitudinal defects, API grade, and wall thickness verification.

The remainder of the wellhead assembly can be installed and tested (the Christmas tree), and now the tubing (which is full of brine) can be displaced to a predetermined depth. Displacement depth is determined by the operator's desire to perforate underbalanced, balanced, or over-balanced. The well can be perforated by using a perforating gun run in on an electric wireline unit inside the tubing and through the packer to the determined depth.

Either the well will flow after perforation, or it may have to be brought in by swabbing or gas displacement. Gas displacement is usually done by introducing nitrogen gas through a continuous 1-in. coil tubing that is run into the tubing to the perforations. When the well has been allowed to flow for several hours, it is clean enough for the drilling department to turn it over to the production or storage department. The rig may now be released and the location cleaned up.

Coring, Logging, and Testing

The pore geometry of a storage stratum is important for determining fluid storage. Other rock properties such as porosity and permeability can be closely related to pore geometry and can be determined by the testing of cores cut from the hole. Cores are rock samples cut from the interval of interest during or immediately after the drilling operation. These cores are tested to determine various points of interest; the most important are porosity, permeability, and water saturation. Permeability needs to be characterized both vertically and horizontally, for the fluid must move in, down, and up toward the wellbore.

Permeability is not the same for liquid and gas, especially at low pressures. Gas moves more easily through rock at lower pressures but approaches the same permeability as liquids at higher pressures. This relationship is best expressed by the Klinkenburg effect of permeability, the formula for which is

$$b=0.777K_{liq}^{-0.39}$$

Eq. 14

where:

b = Klinkenburg permeability factor
K_{liq} = permeability of liquid

Coring is also used to determine the lithology of a stratum, especially when an impervious and impermeable trap is being sought. Aquifer and mined cavern storage projects are good examples.

The crystal geometry of a salt plug can be determined by coring; the purity of the salt can be obtained after a core sample is analyzed. Special care must be exercised in analyzing a core from a salt plug intrusion. The cracking that takes place after salt reaches a lower pressure indicates a high degree of permeability, but the extreme pressures encountered in a salt dome keep the salt crystals from cracking. The usual absence of permeability at great depths makes salt dome cavern storage attractive.

Core samples, although informative, do not indicate everything the storage engineer needs to know about the storage stratum. Logs are run to gather further information about the potential production and storage sands.

Electric logs are used to measure the spontaneous potential to determine the conductivity or resistivity of a sand.

Neutron logging is an applicable means of determining the presence of gas outside the casing. Depending on the lithology and related conditions of a particular field, gas behind the casing could represent migration from the reservoir through the annulus between the casing and the borehole. A neutron log or other suitable log can be run in a well shortly after completion as a comparison reference for future logs.

Temperature logs supplemented by tracer surveys and acoustical logs are useful tools for monitoring possible leakage from storage wells. Where surface casing annulus pressures can be monitored, an early indication of possible subsurface casing leaks or wellhead seal leaks may be detected.

Cement tops and bond condition can be verified by running appropriate logs or by test as required by state regulatory bodies.

A cement bond log or other required state test can be run on production casing to determine the integrity of the cement fill. If cement is found to be inadequate, then suitable corrective measures would typically be taken to ensure the integrity of the annular fill.

Well location sites should be of adequate size and shape to service and maintain all surface equipment installed at the location.

Each wellhead valve should be inspected, serviced, and operated at regular intervals to ensure proper functioning. A record of these inspections should be maintained.

Completion

In designing the actual cementing of the casing string, the operator should recognize the existing subsurface formation and hole conditions such as fluids, sloughing, and temperatures. Then, taking the foregoing into consideration, the operator should consider:

- Designing the proper type of cement, quantity of volumetric fillup, and additives required to ensure good quality of cement completion, recognizing the API Cementing Specifications under Standard 1-A (see Appendix A).
- Ensuring, in so far as possible, before the cementing operation, that all equipment, material, and personnel necessary to complete the operation are on the job location and prepared to complete the operation.
- Using a high-quality cement with good compressive strength, typically 500-psi (3 447.38-kPa) compressive strength or greater.
- Analyzing water used in cementing operations to determine the quality so as to ensure the integrity of cement compatability.
- Determining cement requirements for adequate annular cementing fillup based on factors such as experience, drilling operations, and mud returns. However, consideration should be given to running a hole caliper log to ensure that the cementing operation can be designed properly.

Experience and research have demonstrated that pipe movement during the cement displacement may assist in attaining a good cement bond; therefore the following recommendations can be considered:

- The surface end of the casing string can be installed so as to allow the operator to choose between rotating or reciprocating the casing string. Cementing lines should be secured as much as possible to eliminate danger in case a rupture occurs during the cementing operation.
- Provisions should be made to ensure accessibility to the equipment installed to house the various cementing plugs.
- One or more plugs should be used to efficiently separate the cement from the displacement medium. The plugs should be designed to adequately wipe the casing clean during displacement.
- Excess returns should be used when a well is to be cemented to the surface.
- Various cementing equipment accessories are available to assist

the operator in acquiring a cement bond with integrity. This equipment should be used in accordance with recognized standards and selected in accordance with proven experience.

- Centralizers should be placed on the casing; spacing depends on subsurface conditions. The pipe should be centralized in the hole irrespective of changes in hole direction or deviation from the vertical. Approximately 300 ft (91.44 m) of casing should be properly centralized through the caprock interval and/or the storage reservoir interval. In addition, known critical areas above the storage reservoir should also be centralized. API specifications for casing centralizers are contained in Standard #10 "D" (see Appendix A).
- Staging tools should be considered any time the hydrostatic head exceeds the fracturing pressure of the storage reservoir or when zones of high permeability exist to cause lost circulation. Consideration should also be given to the use of lightweight cement in lieu of staging.
- The total cementing time allowed for the initial cement setup should be completed in accordance with API standard thickening time.
- When the total cementing time allowed for the initial cement setup in accordance with API standard thickening time is less than the total cement job time, then each stage should be allowed sufficient setup time before starting the next stage to minimize lack of cement bond. Circulation should be maintained in the upper stages throughout the waiting time of the lower stages.
- To prevent excessive cement flowback, various devices are available in the form of float equipment. This float equipment should be installed and used in a way that ensures a limited amount of cement being left in the base of the casing string.
- In some instances scratchers or other mud cake removal methods may contribute to a better formation cement bond. If these methods are used, they should be used in accordance with the manufacturer's recommended procedure.

Safety Devices

Circumstances may warrant the evaluation of the use of down-hole safety shut-off valves. A number of subsurface shut-off valves are available that automatically shut in a well when surface control equipment is either damaged or operated in a fashion that permits uncontrolled flow of gas from the well. The choice of a particular type of device depends on the operator's mode of operation, down-hole conditions, and other factors that the operator may consider necessary or desirable.

The consideration that a device be installed could be governed by population density, proximity to highways or multiple occupancy buildings, and the open-flow potential of the well.

A variety of wellhead configurations may be used to suit the operator's mode of operation for a specific field, including wellhead safety shut-off valves, remotely operated valves, and other such equipment that may be deemed prudent or desirable. Such equipment is typically designed, constructed, and tested in accordance with the latest edition of API Specifications (see Appendix A). The maximum working pressure rating should be equivalent to or greater than the maximum pressure that may be encountered in operating the field. Irrespective of other control equipment that may be used in a particular wellhead arrangement, one or more manually operated valves can be installed on the wellhead in such a position that they would shut in and isolate the well from all other surface equipment.

After sufficient time for polarization after the completion of a well, a current requirement survey should be considered to determine if cathodic protection of the well casing is required. A potential survey can also be made over the complete length of the production casing to ascertain if current reversals occur. If the surveys indicate that cathodic protection is appropriate, then a cathodic protection system can be installed. After installation, surveys determine that protective voltages are being maintained. The pipe-to-soil potentials and current flow should be noted in the permanent records of the well.

SURFACE FACILITIES

The storage reservoir and associated well components are required for the storage of natural gas. The storage process, however, must also be supported by surface equipment. Following is a brief description of typical surface facilities required.

Pressure Regulation

Because the pressure limits of some gas storage reservoirs exceed the pressure of the transmission pipeline system serving the storage reservoir, regulators must be installed to reduce the gas pressure to that of the pipeline. Strategically located regulators allow for gas expansion and permit lighter piping to be used. Temperature changes associated with gas expansion must be considered, and, therefore, treatment with methanol or heat to prevent freeze-ups is common.

Measurement

An accurate gas measurement system is important for storage reservoir monitoring and different types of measurement techniques have different degrees of accuracy. Positive displacement meters or orifice meters can be used, and the most accurate method should be evaluated and ultimately installed. Turbulent flow associated with pipe fittings must be controlled with straightening vanes or long meter runs to ensure accuracy.

Compressor Stations

Compressor stations are used to compress the storage gas for entry into the transmission line, the reservoir, or both, depending on the working pressure of the storage system. Stations associated with storage fields usually are concerned only with the movement of gas into or out of the reservoir. When gas must be compressed during withdrawal, the compressor station must raise the pressure above that of the main transmission line. Conversely, when the main transmission line is lower than the storage reservoir, gas must be compressed during injection. Compressor stations are designed for both contingencies. The surrounding environment must be considered, especially in populated areas. Refer to the GEOP series Volume II, *Compressor Station Operations* for more detailed information.

Field Piping System

The field piping system is a series of interconnecting pipe used for gas injection and withdrawal. The same system can be used for both, or separate piping systems can be installed for injection and withdrawal.

Gas Conditioning Plant

Transmission quality gas that is injected and then withdrawn from storage fields becomes contaminated with impurities that must be removed before the gas is reinjected into a transmission mainline.

The most common impurity encountered is water. Water can cause hydrates, which form on expansion of natural gas. Excessive water is typically removed by gas dehydration systems.

Bacteria are sometimes present and must be treated chemically to prevent further problems.

In some cases, hydrogen sulfide is absorbed from a storage reservoir and "sweetening" must be used to remove the sulfur. Sulfur production from a storage field is usually not large enough for commercial sales and it must be disposed of in an economical and environmentally safe manner.

Heavier petroleum ends and condensate may also need to be removed

and can be sold. Oil is sometimes produced in association with storage
gas and is a valuable commodity.

OPERATIONS

The actual storage process is a series of injection-withdrawal cycles
that must be closely monitored. A brief general description of some
operating practices follows.

INJECTION

Because most storage fields operate at higher pressures than the
pipelines that supply them, the gas needs to be compressed before it
is injected into the reservoir. When gas is injected into the formation
it is literally packed, and the more it is packed the less it will compress.
This phenomenon is called "compressibility" and is referred to as the
compressibility factor or z factor. More information on this subject may
be found in many textbooks on chemistry, physics, and gas engineering.

WITHDRAWAL

When gas is needed for sales it is withdrawn from storage, usually
by free flow. When storage reservoirs operate at lower pressures than
mainline pipelines, the gas must be compressed upon removal; when
they operate at higher pressures and are allowed to free flow, they may
be treated with methanol to prevent freeze-offs.

RECYCLE

All storage fields undergo a series of injection-withdrawal cycles,
whereas some production fields undergo a cycling operation in order
to recover more hydrocarbons. Oil fields and especially condensate fields
are good examples of this process. This cycling process is accomplished
by reinjecting produced gas back into the formation to maintain a high
pressure base; this pressure base prevents a condensate gas from
flashing to a vapor. In a storage field this recycling process is done for
a different reason—seasonal injection and withdrawal—although production of the remaining hydrocarbons is still done in many depleted fields
because of recycling.

SHUT-INS

Storage fields may be shut in during the injection and withdrawal process to obtain pressure measurements for inventory monitoring processes. Reservoir pressure tests are done under shut-in conditions so that pressures may stabilize for accurate readings to be recorded.

TESTING

Most companies run an inventory verification test once or twice a year. Bottom-hole pressure readings are taken during shut-in, and qualitative values can be calculated to verify inventory. Several operating parameters can be established with bottom-hole pressure readings, such as unaccounted-for losses or gains, salt cavern creep, and water influx or efflux. Other test procedures establish casing leaks, tubing leaks, wellhead and valve leaks, caprock leaks, and plugged perforations. Some tests evaluate performance such as deliverability instead of looking for problems. A more detailed discussion can be found in the case studies, or in many of the publications cited in the *Underground Storage Gaseous Fuels: A Bibliography,* available from A.G.A., cat. No. XU0178.

WIRELINE

The "Drilling" and "Completion" sections of this chapter discuss some of the uses of wireline operations. A wireline is simply a wire on a large spool for lowering an instrument into the wellbore to set a tool, such as a perforating gun, packer, down-hole valve, pressure bomb, or logging device. Some companies have their own wireline facilities, but most prefer to use the equipment of a service company.

WELL MAINTENANCE

Many maintenance programs can be classified as well maintenance, whether valve repair or down-hole workovers – for example, repairs to packers, cement squeeze for casing leaks, washing of perforations, remedial acid jobs, valve repair, corrosion inhibition, tubing replacement, and change-out. Well maintenance may be on a large scale, such as a workover, but usually is an established maintenance program for preventing major problems.

REFERENCES

1. Bays, C.A., "Use of Salt Solution Cavities for Underground Storage," *Symposium on Salt*, Northern Ohio Geological Society, Cleveland, 1963, p. 564.

Part 2.
CASE STUDIES

Chapter 1

AQUIFERS

Herscher Dome
Natural Gas Pipeline Company of America
Walter M. Rzepczynski

INTRODUCTION

This chapter covers the development of aquifers suitable for underground storage of gas. The historical background leading to the development of this type of storage is discussed, together with considerations for selecting, designing, and developing aquifer storage. Case study examples are drawn from the Herscher, Illinois, aquifer development project.

The development of aquifer storage reservoirs has become a reasonably good means of providing gas storage to areas of the Midwest. April 1, 1953, marked the completion of the Herscher, Illinois, aquifer gas storage reservoir development, one of the earlier and more successful developments. Its success in spite of numerous problems is a tribute to those who pioneered to make it successful. Whether at today's gas prices new aquifer storage reservoirs can be developed is questionable. It could be that only a few existing aquifer storage reservoirs can still be expanded if some storage capacity still exists. The cushion-to top-gas ratio probably makes it difficult to develop an aquifer to gas storage. Whether the use of a cheaper inert gas for cushion gas could allow further aquifer storage development is debatable. The economics may not allow this to happen but if the need is there, then the economics may not be the only important consideration.

PRELIMINARY PLANNING

The most desirable requirements of a worthy aquifer prospect are:
- An adequate structural feature such as a dome or anticlinal configuration.
- An impermeable layer or caprock to provide a cover for any vertical migration of gas being confined under pressure.
- An aquifer sand, or sometimes carbonate, thick and porous enoughedit and capable of accepting and giving up gas readily under a pressure differential.
- Large lateral extent, to allow the compressibility of large volumes of water, because water has such limits of compressibility (see Fig. 7 for a graph illustrating the compressibility of pure water). These prerequisites have to be determined before or immediately after gas is injected for development of a storage reservoir. In the case of the Herscher field, it was obvious that some difficulty would be encountered in establishing these requirements and assuring that no problems would be experienced once gas was injected at a higher pressure than the original hydrostatic pressure of the water-bearing sand.

SITE SELECTION

The prospective structure must be located close to the intended market area. It should be close to the major transmission line, possibly

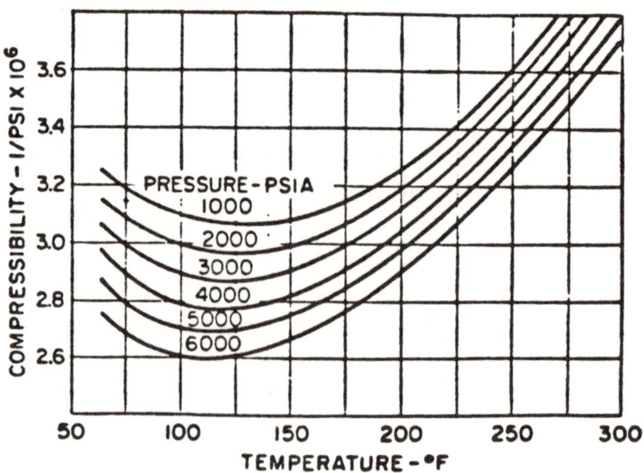

Figure 7. **Compressibility of pure water.**

no more than 10 to 15 mi (1.628×10^4 to 2.443×10^4 m) away. It should, if possible, be away from any populated area and not overlain by any coal mine, other strip mine, or water body.

GEOLOGY

The structure should be identified at the surface, including the information provided by wells already drilled into or through it. Seismic or other geophysical surveys should be used to provide an adequate map of the structural feature that would be used to store the gas (see Fig. 8 for a structural contour map of the Herscher storage field). Any shallow water wells providing geologic marker beds should be used to better define the structural feature. It is important that the structure have a closure (from crest to saddle) of at least 70 ft (21.59 m) and a minimum surface area of at least 1 500 surface acres (6.070×10^6 m²). A larger area is preferable to an area much smaller than 1 500 surface acres. The structure should not be too steeply dipped in any direction or too gently sloped. The ideal depth is at least 1 500 ft (457.2 m) and usually no more than 4 000 ft (1219.2 m), but some aquifers that have been successfully developed are as shallow as 900 ft (274.32 m) from the surface. Drilling permits should be obtained from landowners or leaseholders before drilling.

Once the structure has some identification, additional shallow test holes can be drilled to some representative geologic marker bed to assure a well-defined structural feature. In the direction of the structural nose that has the open-ended last closing contour or saddle, additional wells should be drilled to fully appreciate the closural boundary. If enough wells are drilled, structural faulting could be ruled out based on adequate configuration of the structural map.

Even after the structure is fully mapped on a shallow marker bed, it is possible that as drilling goes deeper to establish the specific aquifer formation, some loss in closure or change in the structural feature could be experienced. Several changes can be found in the location of the structural crest with deeper drilling if the rock layers have been uplifted more than once in geologic time. It is important to have good structural mapping to provide a basis for locating the deeper wells that will be used for edge observation in the water phase after the gas bubble is developed, and as crestal injection-withdrawal wells.

Deeper wells are drilled to the basement rock (usually the granite wash) after the necessary steps have been taken and the structure is mapped. The logs of the first deep well, located at the crest, should be the basis for logging the formations and interpreting the lithology,

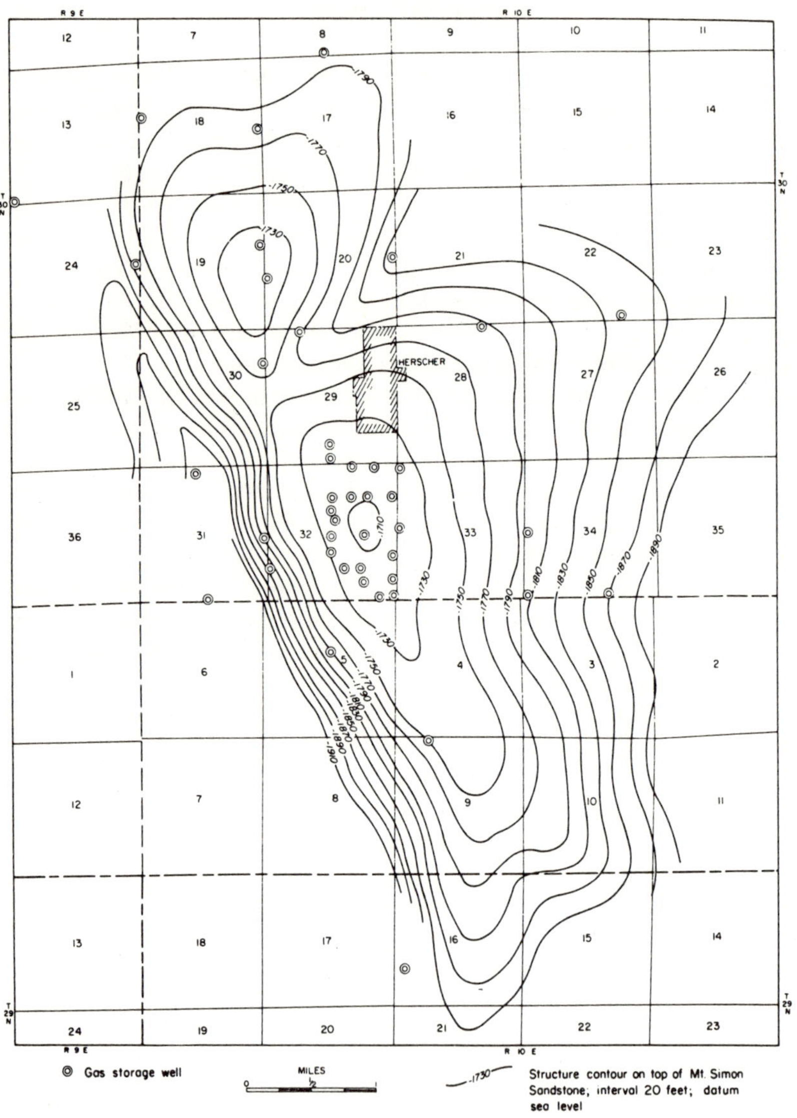

Figure 8. **Structure contour map of Herscher storage field.**

porosity, and density of the various formations. These data are a basis for selecting the aquifers suitable for the storing of gas, providing they are overlain by shales or other dense carbonate rocks that would be impervious to vertical migration of gas. Often the regional geology has already been established by others who have drilled in the area. This knowledge can be used to strengthen the choice of a specific formation for gas storage development.

Once a specific aquifer formation is chosen for gas storage development, it is advisable to drill several deep wells on or at least close to the last closing contour of the structure. The location is important because it can show if closural depth is lost or gained from the shallow, structural, marker bed. These wells can also show if there is continuity of the formation across the entire structure and if any thinning or thickening has taken place. The formations most suitable for potential gas storage development should be cored and the samples analyzed to provide porosity, vertical and horizontal permeability, and threshold pressures of the caprock. An analysis should be made of the irreducible water saturation of the formations into which gas will be injected. This analysis provides the engineer with numerical values of the saturation of the reservoir rock. Porosities of 13 percent to 15 percent are very good, and horizontal Kpermeabilities in excess of 120 mD ($1.8 \times 10ff^{-9}$ cm²) are considered more than adequate.

If several aquifers in the structure look promising for gas storage, the deepest aquifer should be chosen. The upper aquifer can be used as an observation point above the primary caprock of the deeper aquifer. The deeper aquifer would have a greater reservoir pressure, which in turn would mean more storage capacity if the porosity is nearly the same in both aquifers. If the upper aquifer has greater thickness or more horizontal permeability, the upper aquifer should be chosen for gas storage. It is sometimes necessary for each aquifer to be tested by a water "pump test" to fully establish both aquifer reservoir parameters. It could be that both aquifers should be developed, but both should never be started at the same time.

Water samples, with their analyses, may give a better clue as to which aquifer should be developed. Some aquifer waters are contaminated with hydrogen sulfide and may not be suitable for storage development. The water samples sometimes rule out an aquifer even though the reservoir parameters otherwise are favorable for gas storage.

Experience has shown that relatively thin aquifer sands can be used for gas storage. An aquifer sand in Iowa is only 28 ft (8.64 m) thick and has almost been fully developed. After 20 years it is beginning to act like a non-water-drive field, because all the bottom water has been fully drained away.

REGULATORY REQUIREMENTS

The various state commerce commissions authorize gas injection after all storage easements and pipeline rights-of-way have been obtained from the landowners or leaseholders. Condemnation of landowners' property can usually be obtained when it is required. FERC approval is required to withdraw gas from storage reservoirs involved in interstate transmission. Department of Transportation approval is required on all storage fields, interstate and intrastate.

RESERVOIR ENGINEERING

Laboratory tests on cores are normally used to determine the matrix permeability of aquifer zones. However, the formation may have a different effective value than that obtained by averaging core data. Fractures not included in test cores influence the in situ permeability as do other nonuniformities. Pumping tests on a well may be used to obtain the in situ permeability. Along with this pumping, pressure observations on an adjacent well completed in the aquifer provide data for computating the in situ composite compressibility of the water-sand system. Fig. 9 illustrates the location of wells that may be used in a pump test.

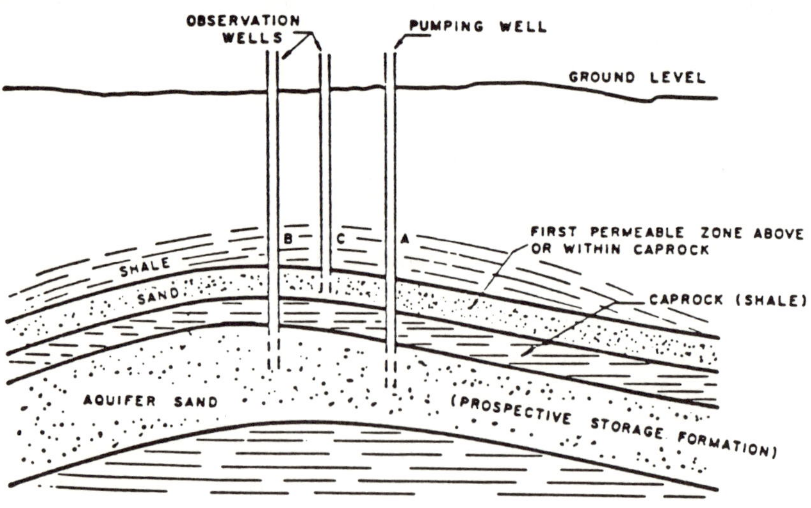

Figure 9. **Pump configuration test.**

The in situ permeability and compressibility of an aquifer can be calculated from the drawdown test on a well pumping at a constant rate. The "point source" solution given by Horner in field units is

$$p = p_o + \frac{70.6\, q\mu}{Kh}\; Ei\left(\frac{-\mu\phi cr_w{}^2}{4(0.00633)Kt}\right) \qquad \text{Eq. 15}$$

where:
  c = fluid compressibility, psi^{-1} (kPa^{-1})
  p_o = initial aquifer pressure, psia (kPa)
  r_w = well radius, ft (m)
   t = time, days
   p = pressure at the well bore at time t, psia (kPa)
   0 = formation porosity, fraction
   K = permeability, millidarcies
   ϕ = fluid viscosity, centipoise
   h = thickness of zone, ft (m)
   q = water pumping rate, bbl/day

$$Ei(-x)^* = -\int_x^a \frac{e^{-u}\, du}{u} = \text{the exponential integral}$$

The in situ permeability and compressibility of an aquifer can be calculated from the drawdown and buildup pressure at an observation well. In Eq. 15, r_w is replaced by r, the distance between the pumping well and the observation well. The pressure of the observation well is given by

$$p = p_o + \frac{162.6\, q\mu}{Kh}\left[\log_{10}\left(\frac{\mu\phi cr^2}{0.00633\,Kt}\right) - 0.351\,3\right] \qquad \text{Eq. 16}$$

where p_o is the initial pressure of the observation well.

An effective compressibility value may be determined by using Eq. 16. If the observation well pressure is plotted as $(p-p_o)$ versus $\log_{10}t$, then the slope of the straight line portion of this curve is $-162.6\ qM/Kh$ and the intercept of this straight line (value of p at $\log_{10}t = \phi$ or $t=1$) is

$$\frac{162.6\, q\mu}{Kh}\left[\log_{10}\left(\frac{\mu\phi cr^2}{0.00633\,Kt}\right) - 0.351\,3\right]$$

---

* Values for Ei (−x) are shown graphically in B. C. Craft and M. F. Hawkins, *Applied Petroleum Reservoir Engineering*, Prentice-Hall, Englewood Cliff, NJ, p. 312. In tabular form, the values appear in Table 1.1, p. 4, of John Lee, *Well Testing*, Society of Petroleum Engineers, Dallas.

from which c may be calculated. Alternatively, any point (p, log t) on
the straight line portion of the curve may be used in the equation

$$p - p_o = \Delta p = \frac{162.6\,q\mu}{Kh}\log_{10}t + \frac{162.6\,q\mu}{Kh}\left[\log_{10}\left(\frac{\mu\phi c r^2}{0.00633\,Kt}\right) - 0.3513\right] \quad \text{Eq. 17}$$

to calculate c.

RESERVOIR SYSTEM

An analysis has been made of unsteady-state water movement in
an infinite aquifer using the formulations and solutions given by Van
Everdingen and Hurst. The assumption was made that water moves to
and from the blanket sand through a radius (R#b) equal to the radius
of the gas reservoir. Although the penetration of the sand by the gas
bubble is only partial at the center and approaches zero at the edge of
the bubble, the derivations assume constant pressure along a-b on the
radius R_b (see Fig. 10 for a schematic of a gas bubble at Herscher storage field). Because the gas bubble is essentially at constant pressure
and the radius of the gas reservoir is small compared to the extent of
the total reservoir, the assumption should not interfere with the basic
nature of the solution to flow problems.

The equation expressing the relationship between pressure, radius,
and time in an aquifer was derived by using the Darcy flow equation,

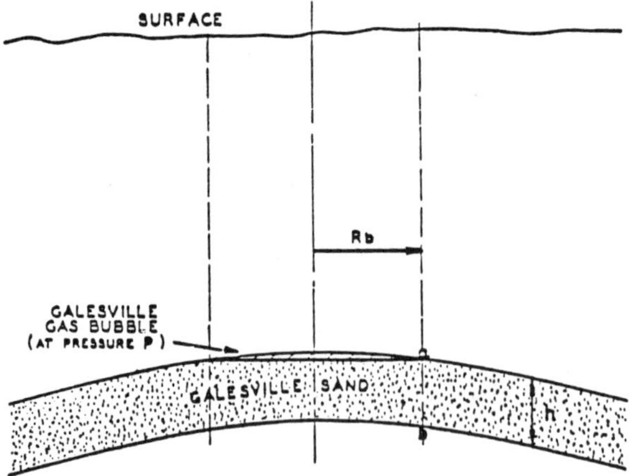

Figure 10. **Schematic of a gas bubble in Galesville sandstone at
Herscher storage field.**

the continuity equation, and a simple expression for the compressibility of the fluid. The difference between the fluid entering and leaving a concentric element of sand was equated to the compressibility of the fluid times the change in pressure to obtain Eq. 18.

$$\frac{(\partial^2 P)}{(\partial^2 r)} + \frac{1}{r}\frac{(\partial P)}{(\partial r)} = \frac{\partial P}{\partial t} \qquad\qquad \text{Eq. 18}$$

where:

   P' = pressure on fluid, atm
   r = radius divided by R_b (dimensionless)
   $t = \dfrac{KT'}{\phi c\,(R'_b)^2 \mu}$, dimensionless time
   K = permeability, darcies
   T' = time, seconds
   ϕ = porosity of sand, fraction
   μ = fluid viscosity, centipoises
   $c = \dfrac{\text{volume}}{(\text{volume})(\text{atm})}$

   fluid and rock compressibility (see Fig. 11)
   R'_b = radius of gas reservoir, cm

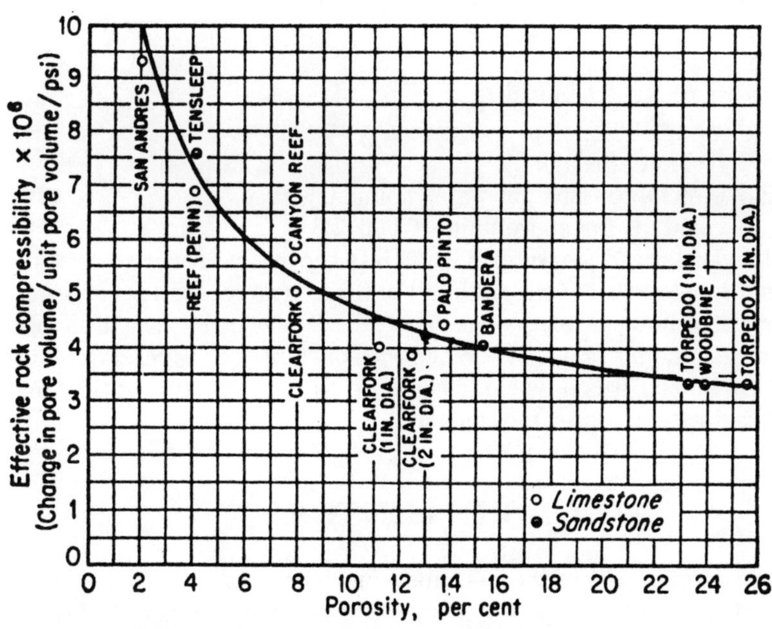

Figure 11. Effective reservoir-rock compressibilities.

By means of the LaPlace transformation, solutions to Eq. 18 were obtained with one variable held constant – pressure or flow rate. These solutions are called "constant terminal pressure" and "constant terminal rate," referring to the pressure and the flow rate at the radius of the gas bubble.

The simplified equation for computing pressure drop at R_b versus time at a constant rate of water movement in radial flow is given by Eq. 19.

$$\Delta P = \frac{(aq\mu)\,P_{(t)}}{Kh} \qquad\qquad\qquad \text{Eq. 19}$$

where:

ΔP= pressure difference, psia

    = (P#2 P#1) when q is water influx

    = (P#1 P#2) when q is water leaving gas reservoir

  a = constant for units indicated # 25 150

  q = thousands of cubic feet of water movement/day

  μ = water viscosity, centipoises

$P_{(t)}$ = function of dimensionless time t,

      where t= $\dfrac{KT'}{\phi\mu c\,R_b{}^2}$

  K = sand permeability, mD

  h = sand thickness, ft

  T = days

  ϕ = porosity, fractional

  $c = \dfrac{\text{volume}}{\text{(volume) (psia)}}$

  R_b = effective radius gas-water boundary, ft

  b = constant for above units = 0.006 326

Eq. 19 was derived for water flowing into the gas reservoir because of pressures below that of the aquifer, but there is no reason for not using it for water flow out of the gas reservoir. Tables of $P_{[t]}$ are available for corresponding values of t. When the constants of the reservoir and fluid are known, it is possible to compute ΔP as a function of time for any fixed rate of water movement.

GAS INVENTORY

The development of the gas bubble and increased inventory volume is directly related to the water efflux (compressibility of water and rock) caused by the delta pressure above the initial virgin aquifer pressure.

The thickness, porosity, and permeability are important to the water efflux rate caused by the differential or delta pressure above the initial pressure acting on the aquifer. Most gas bubbles act in a vertical downward direction, taking advantage of bottom-water drive and not edge-water drive. The degree of vertical pressure transmissibility is proportional to the vertical permeability. If the vertical permeabilities are high in relationship to the horizontal (at least 5 percent of the horizontal), the chances are excellent that the gas bubble growth in the vertical direction will be satisfactory. If the vertical permeabilities are very low, then lateral movement of the gas bubble will be more pronounced, causing undue fingering of the gas bubble if too high a pressure differential is used. An observation well should be drilled below the gas bubble to observe the pressurization of the aquifer below the gas bubble bottom. It is important to know the increase in pressure below the gas bubble in order to know the effective thickness in feet of the aquifer being pressurized. This is required in order to use the Van Everdingen and Hurst unsteady-state equations to obtain the rate of water efflux relative to differential pressure and time. It is also possible to use the Schilthuis steady-state equations and obtain similar results of water efflux. The water efflux rate will then allow computations to be made on the gas bubble inventory growth. Observing the rate of pressure rise in observation wells on the structural edge makes it possible to determine whether the aquifer behavior indicates the aquifer is infinite or finite. In most cases after only a year or two of gas injectivity, this phenomenon can be determined. If the aquifer is finite, it is obvious that even with all other prerequisites of structure, caprock, and good reservoir characteristics, the rate of growth may be limited, drastically reducing the total gas inventory and rate of growth of time. The best aquifer storage reservoirs usually have no more than 125 to 170 psi (861.84 to 1 172 kPa) of pressure differential above the initial aquifer pressure.

The rate of water influx can be established only when gas is withdrawn from the gas reservoir and the reservoir pressure is lowered below the pressure of the surrounding aquifer pressure. This water influx can be determined in the same manner as the water efflux, although doing so is more difficult if the rates of gas withdrawal are held constant.

It probably is best to lower the gas reservoir pressure (150 psi) (1 034.2 kPa) sufficiently below the initial virgin aquifer pressure and then shut in the reservoir and withdraw no gas for a number of days. During this time, if possible, all the withdrawal wells should be shut in so that a good representative pressure can be obtained. The rate of pressure rise in the gas reservoir for a given period ofshut-in time allows one to compute the rate of water influx assuming the measured gas inventory volume has totally remained in place.

Another way to establish the rate of water influx uses data from two water observation wells located in the same direction from the gas bubble and a certain distance apart. As each well exhibits a change in fluid pressure caused by gas being withdrawn from the gas reservoir for a given time period, computations can be made to show the amount of water that is moving in cubic feet per psi per day between wells in the direction of the depressurized gas reservoir. This method allows for determining the distance from the center of the gas reservoir from which no pressure change is taking place. If this maximum distance is known by computations, then the aquifer limit has been identified. If this outer limit can be obtained in all four directions, the aquifer limits or strength for a given delta pressure between the gas reservoir and the surrounding aquifer has then been determined. Thus an appropriate mathematical model can be constructed so that all reservoir predictions can be made over time and inventory change.

As the reservoir grows in gas inventory volume, it is obvious that some gas becomes top storage while the remaining volume becomes base or cushion gas to support a given daily and seasonal withdrawal volume during a cycle of withdrawal from the reservoir. The top storage volume is influenced by reservoir characteristics, market requirements, number of injection-withdrawal wells drilled and completed, horsepower made available, and dehydration facilities. Because of high permeabilities, some reservoirs have high peaking rates of withdrawal. Those with low permeabilities essentially provide base load rates of withdrawal, that is, a constant low rate of withdrawal. In any event, the top storage gas is generally supported by base or cushion gas. Some of this gas is trapped as unrecoverable cushion gas, whereas the other volume is recoverable at some later operating condition, at reduced rates of withdrawal. The quantity of cushion gas that is not recoverable at any time during the life of the field, even if the reservoir is abandoned, is not easily determined. In most cases, no two qualified engineers could determine the same volume of unrecoverable cushion gas any closer than within 2 percent to 4 percent of the inventory. It is believed that the percent of unrecoverable cushion in an aquifer gas storage reservoir is constantly changing with time; therefore no exact number should be applied. It can vary from 27 percent to 52 percent of the total gas inventory from the lowest and highest known values. However, the values do not remain the same forever, and in most cases they go down with time. Because it takes years for gravity segregation to take place in man-time, time is important in determining the volume of the unrecoverable cushion gas volume. The way the gas wells are completed has an important bearing on gas recoverability.

DEFINITION

The most important requisite of an aquifer storage reservoir is an adequate caprock overlying the intended gas bubble to assure containment of the gas with no vertical migration to overlying beds. Naturally, if one is fortunate enough to find an anhydrite (supercap) or shale caprock (very good), a minimal amount of core analysis is required to assure that gas injection can proceed. Some shales have inherent fractures that are not readily noticed until gas is injected. In most cases, shales measure vertical permeabilities of 0.000 007 mD with threshold pressures of 1 000 psi (6 894 kPa). These measurements indicate a very good caprock and an excellent chance that the shale would never allow vertical migration of gas.

Other caprocks such as dense dolomites or limestones have proven to be effective, although a vertical permeability measurement of 0.000 7 mD, with threshold pressures of 400 psi to 500 psi (2 758 to 3 447 kPa) could allow a small leakage of gas through a vertical caprock less than 30 ft thick. When the carbonate rock has "vugs" or fractures, leakage through the caprock is likely. If enough l(lwells are drilled through the caprock, the aquifer's continuity across the structure and its thickness can be ensured, verifying a common depositional time.

Another means of testing the caprock is by conducting a water pump test. A pressure differential is created by producing water from the aquifer below. By observing the fluid level in a well perforated in a porous interval above the caprock, one can detect any pressure response across the caprock. Sometimes it requires weeks of water pumping to develop a significant pressure cone before a response is observed in the completed well above the caprock. This test is worthwhile and should be made if at all possible, even though it tests only a small area of the entire structure. Only a foot or less of fluid response in the observation well could rule out the adequacy of the caprock. The response of the observation well above the primary caprock is good only if it also has an adequate caprock above it.

No matter how much the caprock is tested, unless it is very poor by the above-mentioned indications, injection of gas into the aquifer is usually necessary to prove that the caprock can contain gas at pressure higher than the initial aquifer pressure. For example, at Herscher, two aquifers tested by gas injectivity proved to have poor overlying caprocks after 4 months with a pressure less than 100 psi (700 kPa) above the initial aquifer pressure.

DEVELOPMENT

If the rate of gas inventory growth during the injection cycle is insufficient under normal operating conditions to meet a growing market need, other means should be taken to develop a faster rate of growth. One such method is to remove water from the edge of the gas bubble or from beneath the gas reservoir in the aquifer that monitors the pressure being exerted on it. The number of water removal wells needed to withdraw enough water is directly related to the effectiveness of the water removal that is being acted on by the gas reservoir. It is advisable to drill beneath the gas reservoir to ensure 100 percent effectiveness, assuming that vertical pressure transmissibility is excellent. If too many vertical fractures exist and too much pressure drawdown is required to provide adequate water production rates at the water removal wells, there could be excessive gas fingering. Wells located on the edge of the gas reservoir at least 20 to 30 ft (6.1 to 9.1 m) down the flank from the gas-water contact sometimes are of limited effectiveness. Locating wells on the reservoir edge is a good way to promote faster reservoir development, but this method should be attempted only if the reservoir needs to be improved from a good to a better rate of inventory growth. A reservoir that seems to be growing very slowly may not be materially improved in rate of growth by a supplemental water-removal program. The water removed should be disposed of in another aquifer, either above or below, depending on the relative salinity of the water removed and the water into which it is being injected. This water disposal should not interfere with the gas reservoir operations.

VERIFICATION AND QUANTIFICATION OF LOSSES

It is always important to know whether all the gas injected remains in the gas reservoir itself. Without an inventory verification and monitoring program to assure that no gas is migrating vertically out of the gas reservoir or moving laterally outside the structural control, there is no accountability. It is also a very unsafe practice not to know where all the gas remains in the structure or gas reservoir. Inventory verification is good bookkeeping and ensures the go-ahead in further development of the field.

There is no single way to verify the total inventory of the gas reservoir. Verification is a continuous practice of observing what is happening in the injection-withdrawal wells, on both injection and withdrawal, as well as in the observation wells that monitor gas pressure responses and fluid level changes in the aquifer.

Some companies practice shutting the reservoir in at the end of the gas injection season for 5 to 7 days. During this time, all injection-withdrawal wells are individually shut in and observed each day. Any pressure sinks in a well relative to the neighboring wells should be of concern and fully checked out. In addition, any pressure gradients of the reservoir should be fully understood and evaluated. The rate of pressure falloff over time can be useful in calculating the rate of water efflux or gas inventory volume in place. This same procedure can be done at the end of the withdrawal cycle after a season of gas withdrawal. If the gas reservoir has been sufficiently withdrawn, reservoir pressures could provide a good water influx into the reservoir. Shut-in pressures obtained at the end of the withdrawal cycle can sometimes be more meaningful than at the end of the injection cycle.

The springtime injection period is another good period for observing the reservoir behavior to determine the inventory volume. It is a good practice to use all these methods in a complementary way to ensure that all the gas inventory has been verified.

A good neutron-logging program to determine gas bubble thicknesses and gas-water interfaces can provide a reassurance that all the inventory is in place. Not all the wells need to be logged in the same year for this purpose.

WELL DESIGN

The way injection-withdrawal wells are located or spaced and how they are completed or perforated determine to a large extent the deliverability from a given well or field. It is necessary that early in the life of the field certain well tests be made (particularly relative to sandface drawdowns) to determine flow rates and possible water entrainment in the wellbore. Too high a sandface drawdown can cone water into the wellbore area, causing reduced gas flow rates and eventually drowning out a well. It is advisable to perforate no more than one-half the layer of the projected gas bubble thickness, from the top down, at least with four shots per foot of depth. In most cases 5-1/2 in (14 cm) casing is the ideal production string, although high deliverability rates of flow could use 7-in (17.8-cm) casing to reduce the friction loss in too deep a well. It usually is good practice to test all wells at the beginning of the withdrawal season, sometimes midway and also at the end of the withdrawal season. It is not uncommon to find that some differences occur from year to year caused by limitations of gas injectivity into each well in the preceding summer and shifting of the gas bubble bottom. It is important to recognize these changes to understand and improve each well's potential deliverability.

During the spring, and after the withdrawal season has ended, it is advisable to inject some gas into each individual well just as a matter of practice to dry up the immediate sandface area. In most cases a slightly greater flowing delta pressure should be used to inject into the flank wells because they might be higher in pressure and on the gas-water edge. As the gas bubble becomes pressurized, the better wells will begin to take most of the gas on injection because the gas will be spread more uniformly. Once the gas reservoir is almost fully developed, the gas will go through the best permeable channels so that no unusual pressure gradients will be developed in the gas reservoir itself.

CORING, LOGGING, AND TESTING

A representative water sample should be obtained from the formations of every well completed below and above the primary caprock, before any gas is injected. If any water pump test is conducted from the aquifer below the primary caprock, water samples must be taken both from the pumping well and from the wells observing formations above the caprock. If any fluid communication takes place through the caprock, the samples and analysis might show the mixing of two formation waters based on the analysis before the water pump test. Even after gas injection, it is a good idea to continue analyzing the water above the caprock to assure that no lower formation waters are being forced through the caprock to the formation observed above the caprock. If the waters are saline below the caprock and fresh above, with no disturbance in pressures under a static condition, it is almost certain the analyses are indicating that there is no communication across the caprock.l

COMPLETION

All deep wells drilled and completed through the primary caprock and storage aquifer should be properly cemented to prevent gas leakage. Most injection-withdrawal wells should be drilled and completed before any gas is encountered by the well at the aquifer depth into which gas is being injected. A cement bond log should be obtained to be sure that a good bond exists between the pipe and cement, and between the cement and formation.

TESTING

Before any gas is injected into the aquifer to be developed, all wells should be swabbed and cleaned out and a productivity index obtained so that the in situ permeability of the exposed interval can be determined. Once the wellbore sandface area has been cleaned out and a buildup test has been completed, a static water level should be obtained. This static water level will show whether the aquifer pressure is in communication with the formations above and below, or completely different. (See Fig. 12 for examples of common well placement.) Experience has proved that if observed aquifers in a static condition show different datum water levels and water analysis before any gas injections, in most cases they remain that way throughout the operations of the gas reservoir. Similar waters and datum levels do not particularly reflect communication between the aquifers themselves but sometimes are caused by water pumping 10 to 15 mi (1.6 x 10^4 to 2.4×10^4 m) away.

Even the displacement of the water out of the wellbore itself can be a test if the displacement process is followed with a constant pressure

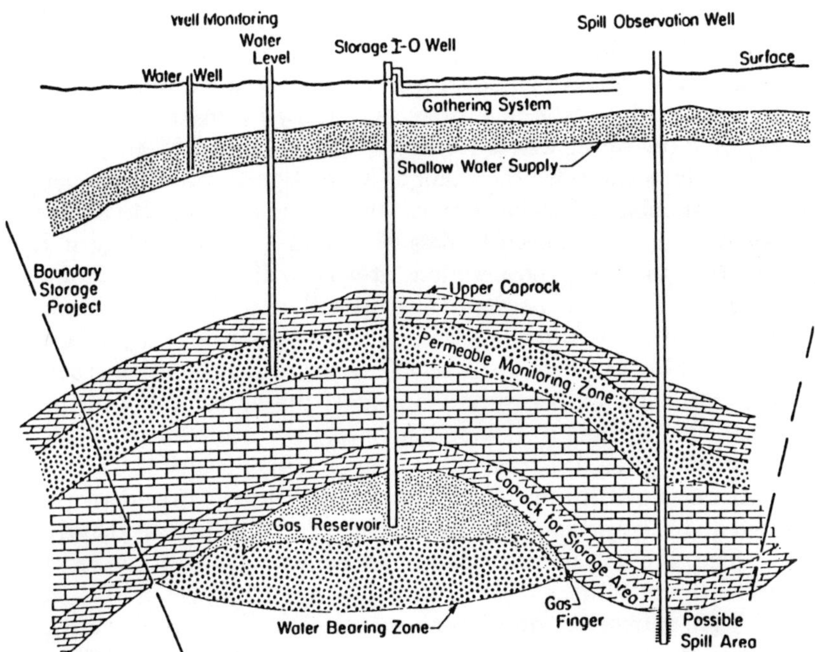

Figure 12. **Schematic showing types of wells used in storage fields.**

differential. It is best to minimize the pressure differential used on this initial period of injection by using 25 to 50 psi (172.4 to 344.7 kPa). There should be a number of injection-withdrawal wells available to displace the water column out of each well before any gas is physically injected into the reservoir itself. In this way, no gas injection well has to use more than the initial pressure used in any of the other injection wells. Once all wells are fully in gas phase from top to bottom, the well pressure should float on a common gathering system. This condition allows each well to begin to form its own gas wafer so that it can gradually coalesce with the adjacent well, thus enabling all wells to form a common gas bubble. (See Fig. 13 for an illustration of the formation of a gas bubble in an aquifer.) If horizontal permeability is high (120 to 200 mD) it may be only a month or two before the wells form one common gas bubble provided the well spacing was 660 or 990 ft (201.2 or 301.8 m). Another factor in formation of a common gas bubble is how the wells have been perforated from the top of the sand or base of the caprock. If only 5 or 10 ft (1.5 or 3.1 m) are opened, the amount of gas fingers is limited and the gas can move away from the immediate sandface faster. It is important to maintain a constant flowing pressure without continuous increasing of the injection pressure with time. The lower the delta pressure in the initial period above the initial aquifer pressure, the better the control of the gas wafers. Caution should be used to get gas into the reservoir while maintaining 25 to 50 psi (172.4 to 344.7 kPa) delta pressure for as long as possible. All water-observation well-fluid levels should be continuously monitored, preferably by charts at each well. The closer the water well is located to the injection-withdrawal wells, the more important are the changes in relation to what is happening in the gas bubble. It has been proved on several pilot tests that vertical leakage from the reservoir takes place almost immediately if it is vulnerable; therefore, observations at each well could be most important. Once a fluid level begins to change with time [1 ft (30.48 cm)], the response with time can be plotted on log-log paper and superimposed on the Non-Leaky Artesian Theis-Curve to see if a deviation is occurring that indicates a pressure anomaly. If the actual curve begins to flatten—that is, fails to rise incrementally with time—it is a sure indicator that a pressure gradient is being broken or is disturbed and that fluid is moving vertically. If the data are properly obtained and interpreted and show no anomaly, the operator can then proceed slowly until at least the initial operating pressures show no leakage even very early in the injection.

All water observation wells above the caprock, like the wells completed in the storage aquifer, should also be continuously monitored. Sometimes a change of only 0.5 ft (15.24 cm) can be meaningful. Once

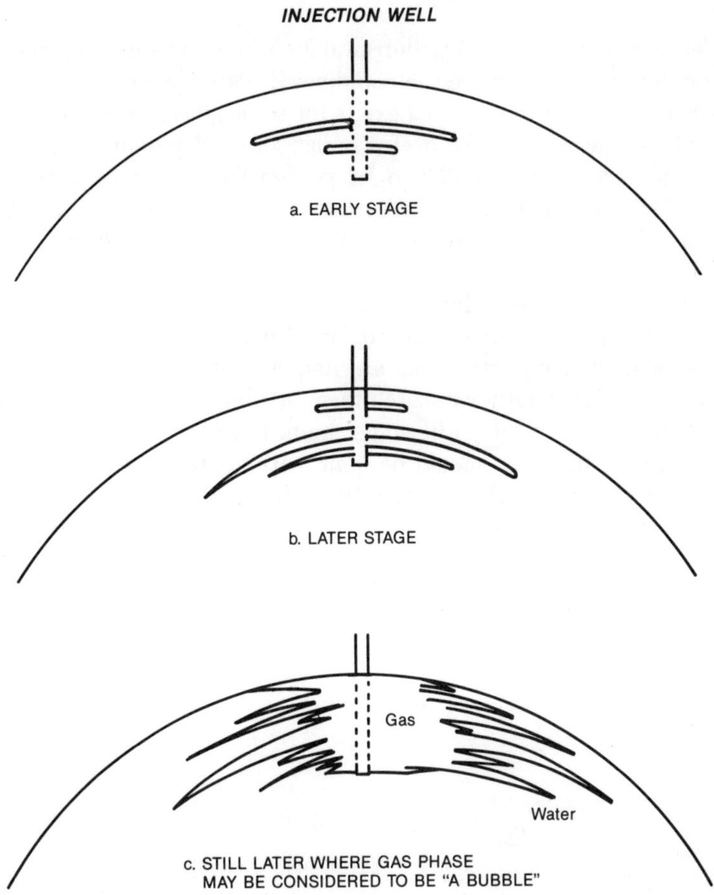

Figure 13. **Formation of a gas bubble in an aquifer.**

the trend is established and the fluid level slowly rises with time, it is then exceedingly helpful to stop the gas injection and see if after several days the rising fluid begins to slow down and fall. Once the gas injection is shut down, the pressure falloff at each well should be observed. In most cases the pressure declines slowly at first and then begins to fall off on a straight-line slope. These individual wells can sometimes show a distinct pressure drop that is a sure sign of leakage near the wellbore. The observation of nothing significant is a good sign and assurance that the wells should be injected into again. These periods of on-injection and off-injection can be repeated until the problem is over.

WELL MAINTENANCE

Because during the gas withdrawal cycle some wet gas is produced and on some occasions free or condensed water is produced in the gathering system, the potential exists for internal corrosion in the low spots of the gathering system. Inasmuch as most pipeline gases contain some carbon dioxide (0.3 to 1.1 percent) when mixed with stagnant water in low spots in the gathering system, pressures greater than 800 psi (5.516×10^3 kPa) can create a mix of carbonic acid potentially capable of causing more severe corrosion than anticipated. It may take years but it can happen. Hydrogen sulfide in the gas coming out of the reservoir is another source of corrosion. Whatever the potential corrosion problem, it is important that an adequate inhibitor be injected into the gathering lines, either in batch form or small daily quantities, that will reach all parts of the gathering system. To ensure that the inhibitor is working, monitoring should be done with coupons or other probes along with a check of the corrosivity of the fluids. The mills per year should never be more than two.

Chapter 2

LEACHED CAVERNS

Valero Transmission, L.P.
Steve Webb

INTRODUCTION

This chapter deals with general principles of design and development of salt cavern leached storage as an alternative means of storing diverse gaseous products and hydrocarbons. The unique characteristics of salt as a means of storage provide high deliverability of gas flow rates with minimal possible losses.

GEOLOGY

Massive salt deposits exist in many locations throughout the world. Many of these salt deposits have been used to construct underground solution-mined caverns for storing such diverse products as natural gas, hydrogen, propane, butane, ethylene, gasoline, and other hydrocarbons.

Salt's unique combination of characteristics make it an ideal formation for cavern construction. It is generally impervious to liquid or gaseous hydrocarbons, has compressive strength comparable to that of concrete, moves plastically to seal fractures or voids, and can be easily mined by dissolution with water.

Three basic conditions must exist if solution mining is to be used for cavern construction:
- Sufficient salt thickness and quality at proper depth
- An adequate supply of raw water for leaching
- An acceptable and economical means of brine disposal

LAND AND RIGHT-OF-WAY

The land (both surface and subsurface) for a salt dome storage system is of three types. First, the land for the storage caverns must be determined from the geologic studies as to placement on the salt dome itself. Once the location is determined, the normal approach is acquisition through purchasing the land in fee. Salt is normally considered a mineral, and the mineral rights must be conveyed to the new owner along with the surface rights. The amount of land is determined by several factors: long-range planning for the number of caverns, type of surface facilities and area needed, and (most important) population density surrounding the project.

Second, the disposal area for brine produced during cavern development is usually some distance from the storage area because salt water formations are normally pinched out across the face of the salt dome. Consequently, geologic studies of the surrounding area are necessary to locate a suitable disposal formation. This land is normally purchased, because only a small area is needed and drilling operations impact most of the surface area for a long time.

The third land acquisition is the water pipelines connecting the storage area to the disposal area. Because no condemnation rights are available for this type of pipeline as in the use of natural gas lines, all pipeline rights must be negotiated with the landowners. Obtaining right-of-way for connecting gas pipeline to the storage project is handled like obtaining a transmission pipeline right-of-way.

REGULATORY REQUIREMENTS

As with any aspect of drilling or construction activity, certain regulatory requirements should be taken into consideration. It is beyond the scope of this book to discuss all of the states' various approval procedures for installing and operating a leached cavern storage facility. However, because much leached cavern storage is in the state of Texas, the federal and state agencies and permits that are involved in construction in Texas are named here.

On the federal level, FERC project approval is required; EPA requires an environmental impact statement and a "prevention of significant deterioration" permit; and the U.S. Army Corps of Engineers requires pipeline permits.

On the state level, the Texas Railroad Commission is the major regulatory agency, requiring permits for organization operating, drilling and completion, pipeline operating, salt water injection, and gas injection. Other requirements are made by the Texas Highway Department (pipeline permits), Texas Water Development Board (casing requirements), Texas Water Quality Board (casing requirements and pipeline permits), and Texas Air Control Board (construction permit).

RESERVOIR SYSTEM

A salt dome facility for storing natural gas does not contain a reservoir system as do other natural gas storage systems such as aquifers and depleted gas or depleted oil reservoirs. A storage cavern is developed in a salt mass below the earth's surface. Therefore the cavern can be treated as a surface tank with walls of varying temperatures. Because the container's (cavern's) salt walls are insoluble in gas and impermeable to gas flow, the only method of gas loss from the cavern is through the connecting pipe to the surface.

GAS INVENTORY

Maintaining accurate cavern volume calculations is very important because the caverns are washed out of a large salt mass containing external stresses and pressures that can cause the caverns to close under minimum pressure. The caverns must be watched closely during the early years of operation to establish trends and operating conditions that minimize maintenance costs and maximize operating availability.

In maintaining storage gas inventories and monitoring for possible losses, the first step is to accurately measure the gas going into and coming out of each cavern as well as the whole facility. This measuring is necessary because each cavern is a sealed system separate from the other caverns.

The next step in gas inventory monitoring is to perform scheduled surveys to determine cavern volume. This surveying is done by bringing the cavern to as close to a pressure and temperature equilibrium state as possible and running a pressure and temperature survey through the interval. With this information and similar information from survey 1 (Fig. 14), a mathematical calculation of the cavern volume can be made.

$$V_c = \frac{\pm V_3}{\left(\dfrac{P_1 T_s Z_s}{P_s T_1 Z_1}\right) - \left(\dfrac{P_2 T_s Z_s}{P_s T_2 Z_2}\right)} \qquad \text{Eq. 20}$$

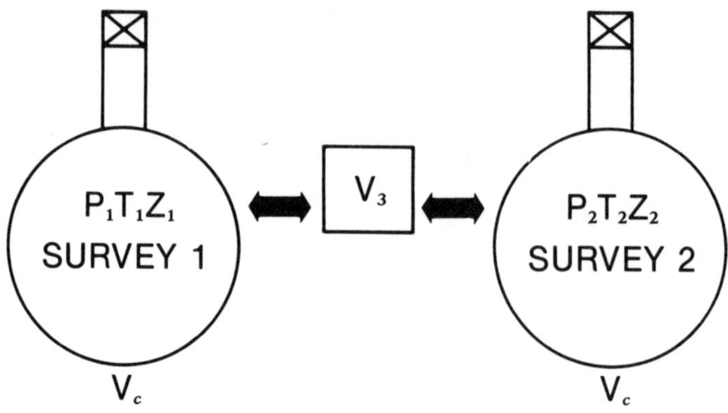

Figure 14. **Survey 1 and 2 data.**

where:

V_c = cavern volume, cf (m³)
V_3 = volume of gas in or out, cf (m³) at standard conditions
    (+ for gas out, −for gas in)
P = average cavern gas pressure, psia (kPa)
T = average cavern gas temperature, °R (K)
Z = average gas super compressibility factor Standard
    conditions: 14.7 psia and 60° F (101.35 kPa and 288.55°K)

where:

s = standard conditions
1 = survey 1
2 = survey 2

This volume is compared to the original volumes determined from the volume of brine removed, the initial volume of gas injected to remove the brine, and the cavern sonar survey taken before removal of the brine. The calculations must be performed on a scheduled basis and as accurately as possible, because any change in cavern size can nullify other monitoring methods.

If gas losses are suspected or determined from these calculations, down-hole logs should be run to determine cement bond, temperature outside pipe, acoustic leaks, and gas charging of water sands (compared to base logs). These logs will help in determining any correction procedures.

DELIVERABILITY

One of the attractive aspects of leached cavern storage is its extremely high deliverability rates. Operating conditions rates generally are so high that, in all of the salt cavern storage facilities, the gas flow has to be choked because gas transmission pipelines cannot handle the high flow rate. Fig. 15 depicts a typical deliverability curve for a solution-mined gas cavern. The durations of these rates are a function of the volume of gas in the cavern. The equation used for calculating the gas deliverability from a cavern is as follows:

$$Q = 155.1 \frac{T_b}{P_b} \left[\frac{p_1^2 - p_2^2 - \left(\dfrac{0.0375(h_2 - h_1)p^2_{avg}}{Z_{avg} T_{avg}} \right)}{G\, T_{avg}\, Z_{avg}\, L} \right]^{0.5} \times D^{2.5} \log_{10}\left(\frac{3.7D}{K_e} \right) \quad \text{Eq. 21}$$

or

$$Q = 17.09 \frac{T_b}{P_b} \left[\frac{p_1^2 - p_2^2 - \left(\dfrac{0.00142(h_2 - h_1)p^2_{avg}}{Z_{avg} T_{avg}} \right)}{G\, T_{avg}\, Z_{avg}\, L} \right]^{0.5} \times D^{2.5} \log_{10}\left(\frac{1.46D}{K_e} \right)$$

where:

$\quad$ Q = gas flow rate to be used in charts, (cf/d at base conditions)
$\quad$ Q_c = gas flow rate to be used in charts, (cf/d at 14.73 psia and 520 °R) (101.35 kPa and 288.55K)
$\quad$ T_b = absolute base temperature (°R–°F+460) (K–°C+273)
$\quad$ P_b = absolute base pressure, psia (kg/cm²)
$\quad$ P_1 = absolute initial pressure (psia=psig+P_a) (P_1+97.9 kPa)
$\quad$ P_2 = absolute final pressure (psia=psig+P_a) (P_2+97.9 kPa)
$\quad$ K_e = effective roughness of inside pipewall, in. (usually assumed to be 0.007 50 for steel pipe) (0.019 05 cm)
Z_{avg} = average compressibility factor at P_{avg} T_{avg} (dimensionless)
$\quad$ G = gas specific gravity (air=1.0)
$\quad$ h_1 = initial elevation, ft (m) above sea level
$\quad$ h_2 = final elevation, ft (m) above sea level
$\quad$ T_1 = initial temperature, °F (°C)
$\quad$ T_2 = final temperature, °F (°C)
$\quad$ D = pipe inside diameter, in. (cm)
$\quad$ L = length, mi. (km)

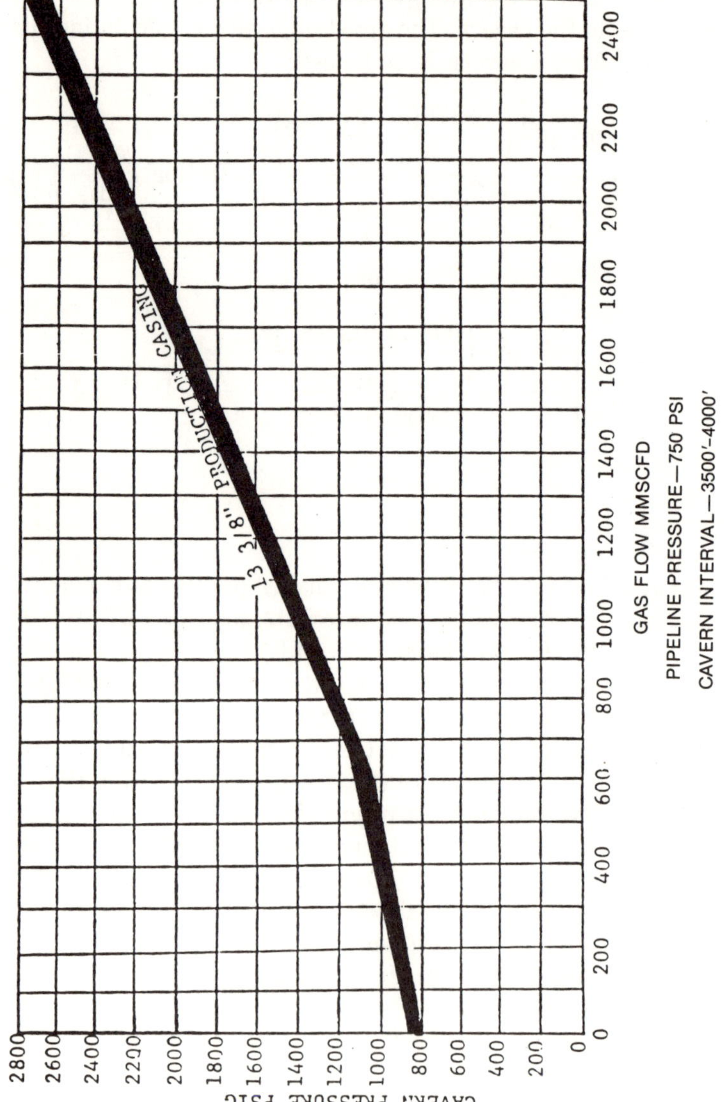

Figure 15. Typical deliverability curve for a solution-mined gas cavern.

$$P_{avg} = \frac{2}{3}\left[P_1 + P_2 - \left(\frac{P_1 P_2}{P_1 + P_2}\right)\right], \text{ psia (kPa)} \qquad \text{Eq. 22}$$

$$T_{avg} = \frac{T_1 + T_2}{2} + 460, \ (^{\circ}R) \text{ where T in } (^{\circ}F) \qquad \text{Eq. 23}$$

or

$$T_{avg} = \frac{T_1 + T_2}{2} + 273, \ (^{\circ}K) \text{ where T in } (^{\circ}C)$$

or

$$\text{Elevation correction term} = \frac{0.0375\,(h_2 - h_1)P^2_{avg}}{Z_{avg}\,T_{avg}}, \text{(psia)}^2 \qquad \text{Eq. 24}$$

or

$$\text{Elevation correction term} = \frac{0.00465\,(h_2 - h_1)P^2_{avg}}{Z_{avg}\,T_{avg}}, \text{(kPa)}^2$$

DRILLING

In most areas, well drilling has already located the salt beds. If the information is inadequate, it may be desirable to core drill the area for a proposed storage cavern; otherwise, the storage well can be cored to sample the caprock and salt section.

In drilling the well, the casing sequences with sizes and the location of the casing seat are most important. That salt dissolves in water-based fluids must be recognized during drilling and cementing operations so that mud and cement properties are not affected.

The surface casing depth and size are dictated by surface waters and normal oil field considerations. The remaining casing sizes are governed by the required flow rates.

The following quotation on well drilling, cementing, and testing, is from Bays[1]:

> Drilling of wells for operation of storage cavities is one requiring an unusual amount of meticulous control during the process so that ultimate certainty of cementing, completion, and development is assured without leakage. Collection of field data, such as drilling time, collection of samples, mud records, and the like, are important phases of the drilling process.
>
> Drilling mud control to prevent undesirable hole enlargement

is very necessary and additives are available to prevent solution enlargement, particularly in the shoe zone where cementing and prevention of leakage are essential. Selection of the casing seat is part of the process of combined field and laboratory study of an exploration hole to assure proper cavity development and protection of roof. This is particularly true in bedded salt; in dome salts it is to be anticipated that the solution cavity will normally extend above the casing shoe either at the well or lateral to it except when the cavity solution process is operated at such rates of extraction as to only permit withdrawal of completely saturated fluid.

As is evident, cementing is an important phase of well construction. Cements used must be impermeable to brine, water, and stored material. They must be nonshrinking in volume as they set. The cement should effectively bond with all the rock types present, particularly in the shoe area and any zone above at which solution conceivably could ever reach the well bore. Bonding of cements has been little studied. In order to obtain the effective cementing needed, an operator may use retarded, sulphate-resistant cements of regular grind, pre-dry-mixed with 1½% to 2½% fine rock salt and 4% to 6% gel drilling mud, this then being thoroughly mixed into a slurry with saturated brine controlled by weight, usually in the range of 14.6 to 16.1 lbs. per gallon (1.8652 to 2.0568 kg/l), to give a slurry that will not shrink or segregate. Such cements are relatively slow in setting and considerably more waiting time is necessary than with conventional oil-well cements.

Drilling-out of the cement shoe or float equipment and any cement left in the pipe can be a critical period in a storage well. The impact of percussion drilling conceivably can create problems as could rough heavyweight rotary drilling. It is desirable to leave as little cement as possible in the well and in most instances to use only a single float-combination guide shoe so as to minimize the drilling-out hazards. If rotary drilling is to be used it is very necessary to weld or use permanent locking compounds on each joint of casing in the lower part of the casing string. Centralization of the casing at a point selected geologically is highly desirable. Centralizers should be located in places in the hole known to be of proper diameter and in strata of high resistivity.

Casing grades selected for storage wells often desirably are of heavier weight than might be conventional for the depth in order to provide more amply for the anticipated length of life,

which no one is yet able to estimate, especially in terms such
as the corrosion of the casing. When the casing is made up, it
is usual for thread dope to be used. Special attention to the
thread-sealing compound is needed in storage cavity wells to in-
sure the insolubility of this material in the stored materials,
brine, or water that may be in prolonged contact with the pipe
during years of operation.

FIELD TESTING

The quotation from Bays[1] continues:

To preserve the stored material and to provide safety, it is essential that the casing and shoe area be completely tight. After cement has set-up and the well is ready for completion, field testing
is desirable. Such field tests should include the data necessary
for reference to later observe physical changes in the cavity and
include both pressure tests and logging done incidental to the
completion process.

CAVERN CONSTRUCTION

The fundamental technique of cavern development is simply to drill
a hole into a salt formation, inject water into the hole, allow time for
the water to dissolve the salt, and displace the resulting brine from the
hole. As the salt dissolves, the hole enlarges, eventually forming a cavern.
In actual practice, the procedure for cavern development is somewhat
more complex and may employ either of two rather different techniques,
"single-wall" cavern construction and "fracture-connected" cavern construction; the single-well technique is most closely associated with construction of caverns in salt domes.

SINGLE-WELL CAVERNS

Single-well caverns offer many advantages and are normally
employed where salt thickness is adequate for caverns of the desired
storage volume. The following discussion outlines the construction of
a typical single-well cavern in a salt dome as shown in Figure 16.

LEACHING PROCESS

When injected into a salt cavern, both fresh and brackish raw water

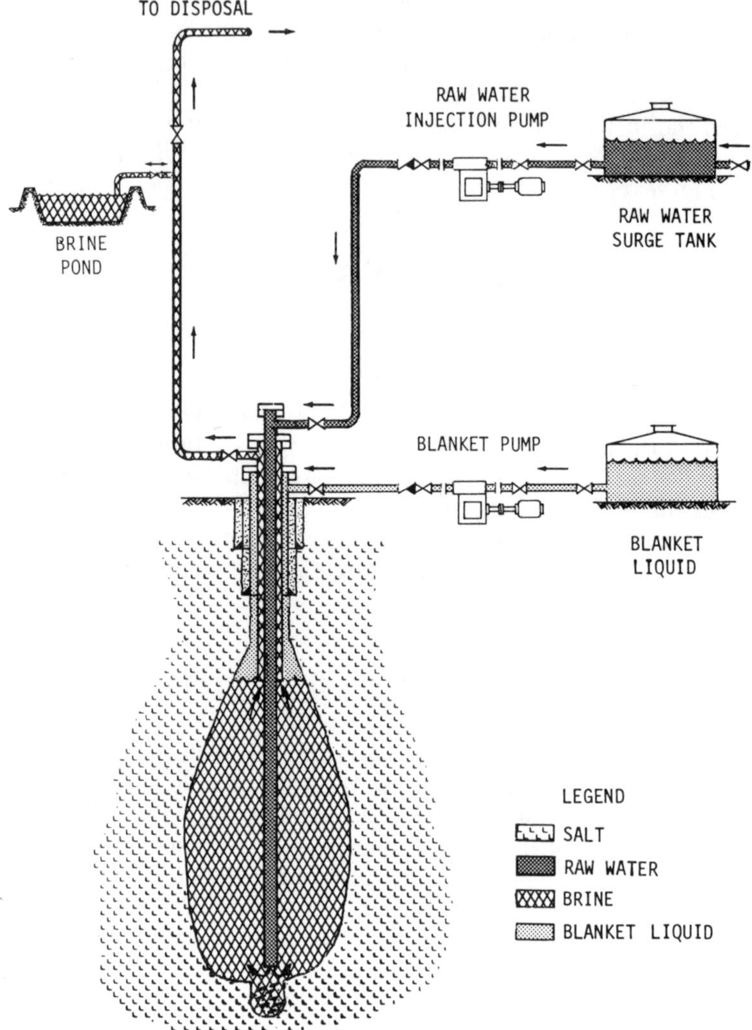

Figure 16. **Construction of a typical single-well cavern in a salt dome.**

becomes saline because of the circulation of unsaturated fluid to the face of the salt, where dissolution occurs. During the initial injection process, the pressure differential between the point of entry and the point of exit from the cavern forces circulation to occur. As the cavern grows, the injected raw water, which is lighter than brine, tends to rise and causes a convection effect, which provides the necessary circulation during the remainder of the leaching process.

CIRCULATION METHODS

Raw water may be introduced into a cavern by either direct or reverse circulation. Direct circulation is most common, involving injection of raw water near the bottom of the cavern and withdrawal of brine through the casing annulus near the cavern top. With direct circulation, the maximum diameter occurs near the bottom of the cavern and the minimum diameter occurs near the top. This type of cavern is commonly known as a "coke bottle" type.

In reverse circulation, raw water is injected down the casing annulus, entering near the top of the cavern and displacing brine into the tubing at the cavern bottom. Reverse circulation develops a large-diameter roof with the span decreasing toward the bottom, resulting in the so-called morning glory-type cavern. Direct and reverse circulation methods use the same drilling and casing procedures, and both methods may be used to control the cavern configuration.

BLANKET

Blanket material is usually liquid propane or diesel oil injected into the space at the top of the cavern. The blanket, being lightest, floats on the raw water or brine.

The blanket is important and requires careful monitoring and control. Its purpose is to prohibit leaching of the salt from around the cemented casing, thus ensuring a pressure-tight cavern. The blanket protects the cemented casing from corrosion during leaching and also prevents leaching of high spots in the roof of a cavern from which the stored product cannot be retrieved. During initial stages of the injection process, the blanket may be used to depress leaching to the bottom of the borehole in order to construct a sump.

SUMP

Insoluble material is present to some degree in most salt. As leaching proceeds, insoluble material accumulates in the bottom of the cavern and may plug the wash casing. This condition can be prevented if a small cavern is first leached below the location of the main storage cavern. The wash casing is then raised to leach the main storage cavern and insoluble material is permitted to accumulate in the sump.

BRINE DISPOSAL SYSTEM

A subsurface brine disposal system can be successful only if a porous, permeable formation of wide areal extent is available deep enough to ensure storage of the injected fluids. An impermeable zone, such as shale or evaporate, must overlie the injection horizon to prevent vertical migration of the wastes or displaced formation brines into freshwater aquifers or even to the surface.

The three most important characteristics of a disposal formation are
- High porosity and permeability
- Large disposal formation thickness
- Large areal extent

Formation porosity is the fraction of total volume that is not occupied by the solid framework of reservoir rock. The porosity of a formation is a measure of the fluid-carrying capacity of the rock. Because fluids within a formation are slightly compressible, there should be a large amount of freely connected pore space to allow for compression of the fluid.

Permeability is the property of a porous medium that measures the ease with which fluids flow through the medium under the influence of a driving pressure. The product of the formation permeability and thickness is referred to as the formation capacity. The allowable injection rate is directly proportional to the formation capacity. Therefore the higher the capacity, the higher the injection rate for a given pressure differential.

Because the fluids within the disposal formation must be compressed to allow room for the injected fluids, the formation should have a large areal extent if it is used for disposal purposes for any length of time.

SURFACE FACILITIES

In a salt cavern storage facility, only the friction losses through the production casing and the storage cavern pressure affect the deliverability capabilities; therefore compression on withdrawal is not normally necessary. Also, as previously discussed, a minimum cavern pressure is necessary to prevent cavern closure (salt creep). With this philosophy in mind, the compression facilities can be designed on injection conditions and rates with no consideration given to withdrawal compression.

The depth of the cavern determines the maximum storage pressure. Each gas pipeline has a standard maximum operating pressure, which determines the exit pressure from the facility. The regulation, gas dehydration, and measurement are then designed from the pressure ranges (maximum cavern and minimum pipeline) and flow rates required. The field gathering system is usually very short—measured in feet (meters)—and offers little restriction to gas flow compared to gathering systems for other types of gas storage in which the field gathering systems are normally measured in miles.

Because the salt is neither soluble nor permeable in natural gas, no changes occur to the gas in storage except when it comes in contact with the solution brine or blanket liquid. The absorption of either or both of these materials is removed in the dehydration facilities in the withdrawal cycle before being discharged to the pipeline system.

CAVERN CONVERSION AND OPERATIONS

When the leaching operation has been completed, the cavern is prepared for gas storage in the following manner:
1. The diesel oil blanket material is removed from the cavern.
2. Both the blanket casing and the wash casing are removed.
3. The wellhead used for the leaching operations is replaced with the permanent operating wellhead.
4. The cavern is pressure tested to assure its integrity.
5. A packer was set immediately below the surface and the wellhead is tested separately at a higher pressure.
6. The dewatering string is suspended from the wellhead to the bottom of the cavern.
7. Wellhead valves and automatic controls are installed.

The conversion is then complete and the cavern is ready for gas injection. Compressed gas is injected into the annulus around the dewatering string, and gas pressure forces brine to rise through the dewatering string, thereby evacuating the cavern. The dewatering string can then be stripped out of the hole or left in at the discretion of the operator. Gas injection is continued until maximum design pressure is reached.

Because gas is compressible, the storage capacity of a cavern is directly related to the maximum allowable storage pressure. A small, deep cavern can store as much gas as a large shallow cavern because the amount of pressure that salt can withstand without failure increases as the weight of overburden rock increases.

The cavern operates as a pressure vessel. As gas is added to the cavern, the pressure rises. As gas is withdrawn, the pressure drops. As the pressure in the cavern drops, the gas delivery rate from the cavern drops and eventually would become unacceptably slow. Before that point is reached, however, gas withdrawal is usually halted at a minimum operating pressure selected to minimize the problem of cavern shrinkage. Gas remaining in the cavern is called "cushion gas." The difference between the volume of gas stored at maximum pressure and volume of gas stored at minimum pressure is called the "usable or working storage capacity."

If a deep cavern is operated as a high-low pressure vessel, unacceptable capacity shrinkage may occur. Salt is a plastic rock that deforms under the influence of heat and pressure, both of which increase with depth. Consequently, a pressure reduction tends to induce more rapid salt movement in a deep cavern than in a shallow cavern. Some known instances of shrinkage in deep caverns exceed 30-percent reduction of the original volume. In these cases, far less shrinkage would have occurred if higher cavern pressures had been maintained.

If cavern stability were ignored, the cavern depth would be found by computing the least cost per million cubic feet (cubic meters), considering construction cost and long-term compression cost. Past evaluations of that sort have led to design depths of between 5 000 and 7 000 ft (1 524 and 2 134 m). Unfortunately, caverns constructed at these depths have experienced high shrinkage rates when cycled to low operating pressures.

The laws of rock mechanics necessary to accurately predict plastic behavior of salt under varying cavern conditions are questionable, but some useful rules of thumb have been developed. Experience indicates that for depths of 3 000 to 5 500 ft (914 to 1 676 m) the pressure at the bottom of the cavern can be permitted to fall about 3 000 psi (2.069×10^4 kPa) below the theoretical overburden pressure gradient of 1 psi/ft (19.76 kPa/m). A cavern should be capable of holding as much pressure as is imposed by the overburden, but maximum pressure is usually established at 80 percent of the overburden at the cavern roof to provide a margin of safety. Thus, if a cavern were constructed between 3 500 and 4 000 ft (1 067 to 1 219 m), the operating pressure would range between 700 and 2 800 psi (4 826.5 and 19 306 kPa). If a cavern were constructed from 4 000 to 4 500 ft (1 219 to 1 372 m), the operating pressure would range between 1 200 and 3 200 psi (8 274 and $2.206\,4 \times 10^4$ kPa).

A gas cavern can also be operated with a brine displacement or a freshwater displacement technique – a wet cavern operation – by pumpng brine or water into the cavern as the gas is withdrawn, thereby maintaining a nearly constant cavern pressure and a nearly constant gas deliverability rate. However, the wet cavern operation is normally not used unless it is desirable to remove all the cushion gas or to enlarge the cavern volume (see Fig. 17 for a comparison of operating techniques for dry and wet caverns). The advantages of a dry cavern operation are that it requires

- No construction of a large, expensive brine storage reservoir
- No brine displacement casing (which lowers the maximum deliverability rate) in the storage well
- No production of raw water or brine displacement
- Less compression energy than a wet operated cavern

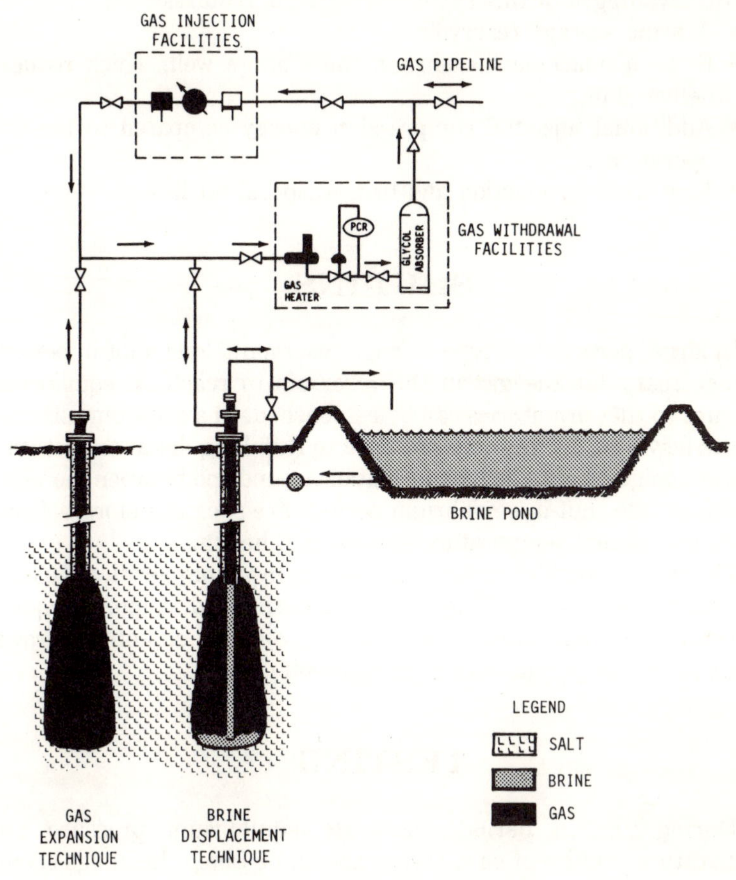

Figure 17. **Comparison of operating techniques for dry and wet caverns.**

The disadvantages of this system are that
- The volume of cushion gas is unavailable for use.
- There is less cavern stability at the minimum operating pressure.
- The gas delivery rate varies.

The advantages of a wet cavern operation are that it
- Maintains a high degree of cavern stability because it is always under pressure caused by a full head of brine or fresh water
- Maintains a constant gas delivery rate

- Allows recovery of all the gas
- Provides for cavern growth if the gas is displaced with fresh water

The disadvantages of this system are that it requires

- A brine storage reservoir
- Brine displacement casing in the storage well, which reduces deliverability
- Additional injection compression energy compared to the dry operation
- Raw water-production and brine-disposal facilities.

SHUT-INS

In large, porous rock-type storage reservoirs, long shut-in periods are necessary for the gas in the reservoir to reach an equilibrium pressure so that inventories can be established. In a salt dome storage cavern, however, the gas does not have to flow long distances between wells and migrate into the rock formation around and between the wells. Therefore, the shut-in equilibrium periods are shorter before bottom-hole pressure and temperature surveys can be run.

The shut-in equilibrium tests are used in inventory verifications and cavern size calculations. They should be performed at more frequent intervals than the standard end-of-injection and end-of-withdrawal surveys used on porous rock-type reservoirs.

TESTING

During shut-in periods, accurate equilibrium pressure and temperature profiles of each cavern should be taken. Readings should be taken in a step-by-step survey through the cavern interval so that an average cavern temperature and pressure can be calculated mathematically. Knowing the average pressure and temperature throughout the cavern interval is necessary to obtain accurate gas inventory and cavern size calculations.

REFERENCES

1. Bays, C.A.. "Use of Salt Solution Cavities for Underground Storage," *Symposium on Salt*, Northern Ohio Geological Society, Cleveland, 1963, p. 564.

Chapter 3

MINED CAVERNS

Leyden Mine Storage
Western Gas Supply Company
W. Kenneth Hooker, Jr.

INTRODUCTION

This chapter describes the selection, development, and operation of the only underground storage facility in North America currently using an abandoned coal mine to store natural gas—the Leyden Mine storage facility near Denver, Colorado. There are only two others in the free world, both in Belgium.

During the late 1950s, a Colorado public service company determined that a natural gas storage project would be the most feasible way to optimize natural gas supplies. The first step in planning was to retain a consulting firm to prepare a comprehensive geological and engineering investigation of potential underground storage sites in the Denver area. The geographic parameter for their search was the area described by a 25-mi (40.23-km) radius from downtown Denver, including a 40-mi (64.36-km) oblong area toward the northeast.

By October 1958, the consulting firm had completed its search. After review of 16 potential underground storage sites, it was decided to test and evaluate the site known as the Leyden Lignite Mine. The mine had been abandoned since 1950 and was located approximately 14 mi (22.53-km) northwest of downtown Denver. In addition to meeting geographic criteria, the mine location was perfect; it was located at a point in the distribution system farthest removed from major pipeline purchase points.

GEOLOGY

The formations that are exposed in the Leyden area are Late Cretaceous in age, including the Pierre Shale, the Fox Hills Sandstone, and the Laramie and Arapahoe Formations. The following description of the geology in this area was prepared by Robert M. Meddles, supervisor of development of Fuel Resources Development Company.

The mined coal seams are located in the lower 200 ft (60.96 m) of the Laramie Formation, which is 500 ft (152.4 m) to 700 ft (213.36 m) thick in the area of the Leyden Mine and is made up of a sandstone-shale sequence lying between the top of the Fox Hills and the base of the Arapahoe. In 1957 Richard Van Horn described the Laramie and Arapahoe Formations as follows:[1]

> The Laramie is composed of light to medium-gray, medium to fine-grained quartose sandstone, light to medium-gray claystone, and some coal in the lower 200 feet (60.96 m). The coal is sub-bituminous and generally occurs in discontinuous beds. The numerous fossil-leaf fragments that occur in the formation indicate the terrestrial origin of the Laramie Formation. The Arapahoe Formation over-lies the Laramie and at the base is a thick discontinuous conglomerate. The top is marked by a change in lithology from quartzose sandstone and a claystone of the Arapahoe, to tufaceous sediments of the Denver Formation.

Arapahoe Formation in the Leyden Mine area is present in part of Sec. 34, the SE ¼ SE ¼ of Sec. 37 and the S ½ S ½ of Sec. 26, T2S, R70W. The basal conglomerate of the Arapahoe is absent in the Leyden Mine area.

Leyden Creek travels from west to east through Secs. 28, 27, and 26 and beyond from its source west of the hogback. The creek is in a small valley bordered on the north and south by flat-lying pediment surfaces that are covered with coarse gravel. The north pediment is called Rocky Flats; the south pediment has no name and is almost 200 ft (60.96 m) lower than the Rocky Flats pediment.

Two seams of coal in the Laramie Formation have been mined in the Leyden area. Coal mined from shafts No. 1 and No. 2 is from the A seam, whereas coal from shaft No. 3 is from the B seam. The A seam is approximately 50 ft (15.24 m) higher than the B seam, and is almost absent near shaft No. 3. This absence was not known until 1913, when shaft No. 3 was sunk. The shaft passed through a thin coal seam and about 25 ft (7.62 m) lower struck another coal seam. At the time, miners assumed that the lower coal was the A seam. However, an entry driven east from shaft No. 3 failed to enter the A seam workings. A connection was finally made vertically to the A seam workings for ventilation. Later this connection was extended to the surface for use as an air shaft. This shaft is the only known communication between A seam and B seam (see Fig. 18).

The coal seams have a gentle dip of 150 ft/mi (28.41 m/km), or 2°

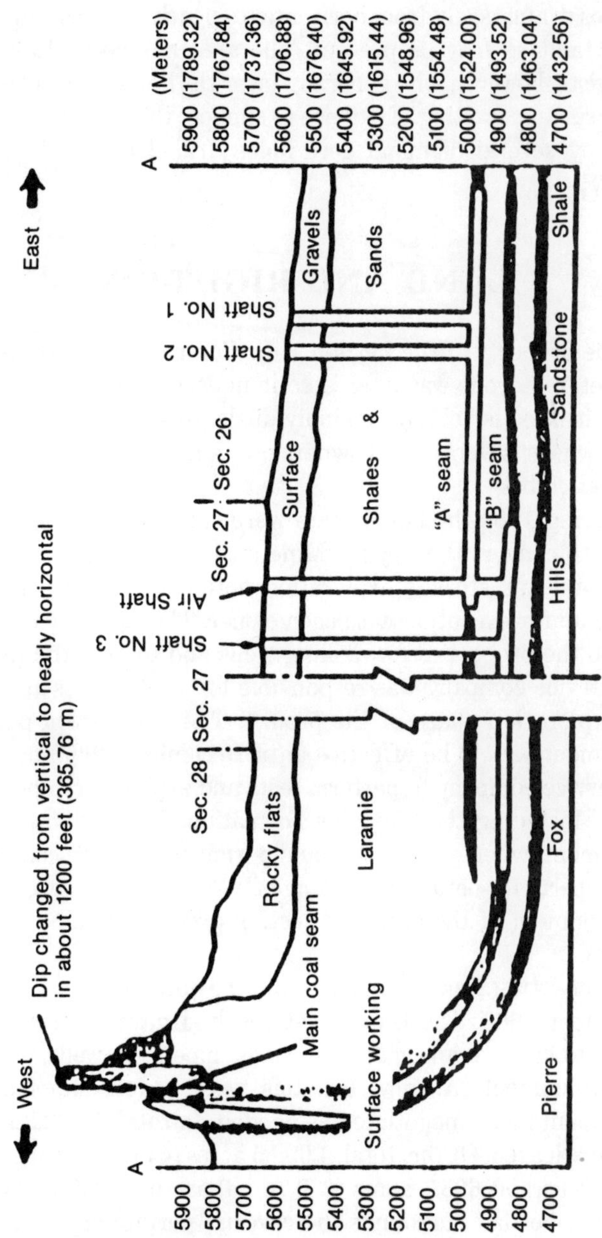

Figure 18. Cross section of A and B coal seams and mined shafts of the Leyden Mine.

to the southwest. A cross section from the east to the west almost
parallels the dip (see Fig. 17). The A and B seams are nearly horizontal
to the western limits of the workings. About 1 000 ft (304.8 m) west
of the limit of the western workings in shaft No. 3, the dip of the beds
changes dramatically from horizontal to nearly vertical, suggesting that
major faulting may be present. The area has been studied by many
geologists, however, who generally favor the bending concept with some
minor surface faulting. Evidence to support the bending concept is provided by several diamond core holes drilled by the U.S. Geological
Survey[2].

LAND AND RIGHT-OF-WAY

The property of the Leyden Mine Company was sold at public auc-
tion before anyone was interested in underground storage. The majority of the land is held by two individuals. In addition to these two principal owners, several small owners own a mixture of surface rights and
mineral rights.

During 1959 the landowners were contacted and informed of the
desire to convert the Leyden Mine into a natural gas storage cavern.
A natural gas storage exploration option and lease agreement was
drawn up for execution with each owner. This agreement provided access to the properties for drilling wells and testing the Leyden Mine.
The service company was responsible for all damages associated with
the exploration phase of the project. The exploration portion of the
agreement was to be effective until December 1960 and would allow
the service company to perform tests and exploratory operations down
to and including the Fox Hills Formation. The option portion of this
agreement, when exercised, would permit the service company to lease
the property for natural gas storage for 25 years. The agreement allowed
abandonment of the project if testing proved the mine unsuitable for
storage.

The testing proved favorable and the options were exercised before
December 1960. The leases that resulted gave the service company
rights to operate and maintain roads, pipelines, wells, and other surface equipment. As new facilities were added, supplemental lease
agreements were negotiated as needed. In total, 1 792.2 acres (725.29
ha) were leased. Of this total, 1 098.5 acres (444.55 ha) are directly over
the cavern and 693.7 acres (280.74 ha) are in a 200-ft (60.96-m)-wide
buffer zone that surrounds the cavern's perimeter.

REGULATORY REQUIREMENTS

The Colorado Oil and Gas Conservation Commission required letter notification of the intent for the initial evaluation of the Leyden Mine. Once it was decided to develop Leyden into a storage project, it was necessary to file an application with the commission. A hearing was then held to determine the interests of the various parties involved. After approval was granted, it was necessary to routinely report drilling and injection activity. After operations began, the commission required an annual report detailing injection and withdrawal activity.

The Colorado Mine Inspection Department informed the service company that the planned work did not fall within the department's jurisdiction and no special permit was required. Based on advice from the department, a 200-ft (60.96-m) buffer zone around the mine's perimeter was included in the applicable leases.

In any case, where a water well is drilled or water is taken from a withdrawal-injection well, it is necessary to obtain a permit from the State Water Resources Engineer.

Public Law 93–275, starting in January 1975, required all companies that are operating underground storage fields and are not subject to FERC [then Federal Power Commission (FPC)] jurisdiction to regularly report injection-withdrawal activity. This information is reported monthly to the Federal Energy Administration.

RESERVOIR SYSTEM

The Leyden Cavern was formed by mining operations that removed approximately 150 MMcf (4.24×10^6 m^3) of lignite from an 8-ft (2.44-m) seam about 800 ft (243.84 m) below the surface. The four shafts used during the original mining operation were sealed to obtain a gas-tight cavern. This cavity now stores 2.6 Bcf (7.36×10^7 m^3) of natural gas at 250 psig (1 825.3 kPa). Pressure changes occur during shut-in periods and are attributed to gas migrating in and out of the surrounding formations.

The walls of the cavern are composed of lignite coal, and the roof and floor are made up of various degrees of sandstone and shale. In areas where shale was present, the miners were forced to leave closely spaced supports to minimize roof caving and floor heaving. The sandstone sections were mined leaving much wider areas.

At the beginning of this project, the mine workings were full of water that had to be removed. Removal was done by drilling wells into the highest points of the cavern and injecting compressed air. At the cavern's depth it was reasoned that a hydrostatic head of 700 ft (213.4 m) would hold about 300 psig (2 170.0 kPa).

After it had been determined that the mine was suitable for natural gas storage and the shafts were sealed, a system of injection-withdrawal wells was drilled into the cavern. The objective was to complete these wells in the void haulage ways left by the original miners, thus limiting pressure drop to tubing and surface facilities.

GAS INVENTORY

A portion of the gas that migrates into the surrounding rock at high pressure is recoverable when the cavern pressure is lowered but does not contribute pressure support during short periods of high withdrawal. Thus the combined volume of cavern and rock storage is used for inventory, whereas the deliverability is determined by the amount of gas in the cavern only.

To measure the total storage volume, an initial stabilized pressure is first determined, which requires an actual shut-in period of about 150 days. Empirical equations have been developed to predict a stabilized pressure based on the initial shut-in pressure. A substantial injection or withdrawal period is then metered and a second shut-in pressure is again converted to a stabilized pressure. These data are used in a simple volumetric relationship to determine the actual amount of gas in place. The accumulated total of gas injected and withdrawn since the last inventory adjustment is totalled with the last adjusted figure for storage volume. These two figures are compared and the difference is considered an unaccountable loss.

DELIVERABILITY

The Leyden storage project has been developed in phases since 1960. The most recent injection-withdrawal well was completed in 1976, bringing the gas well total to 13. The wells vary greatly in deliverability for a number of reasons. The individual well deliverabilities range from 2.6 MMcf/d (7.362×10^4 m3/d) to a maximum of 55 MMcf/d (1.557×10^6 m³/d).

The better wells are the ones drilled into the main haulage ways of the mine, particularly those in areas of sound roof conditions. The best well, No. 8, was drilled into the "mule barn," a large room where equipment and animals had been housed in the mine near one of the shafts. The poorer wells were drilled into areas that sloughed in, in depressions where water and mud collect, and support pillars left by the miners. In the last case, hydraulic fracturing was used to break into the cavern with some success.

Deliverability of each well and the total system is normally tested in the spring each year. The beginning cavern pressure is checked and the wells are brought on individually for a multipoint back-pressure test. The test results are compared with results from previous years to determine workover requirements for that summer.

The total field deliverability is defined by a multipoint back pressure test between the cavern and the suction header of the compressor station with all wells and gathering lines operating. The current rated capacity of the project is 185 MMcf/d (5.239×10^6 m³/d). This rating takes into account the field, the gas conditioning plant, the compressor station, and transmission lines. The 1980 performance test indicated that the field could deliver 185 MMcf/d (5.239×10^6 m³/d) with 125-psig (963.4 kPa) in the cavern. The project could then deliver gas for 5 continuous days at the rated capacity, starting with a 250-psig (1 825.3-kPa) cavern pressure.

SHAFT SEALING, DRILLING, CORING, AND TESTING

Work on the test phase of this storage program was begun on April 17, 1959, with the drilling of the Leyden No. 1 well. During the drilling operation, the last 200 ft (60.96 m) were core drilled to determine the nature of the caprock immediately over the mine. These cores were encouraging, indicating that although some of the strata over the mine were porous sandstone, there were enough layers and depths of tight clay shales to provide an impervious roof. However, this well bottomed out in an unmined area rather than in the cavern as was expected. The accurate locations of coal pillars left in place to support the ceiling of the mine were not available, and it is presumed that this well ended in one of these pillars.

It was immediately decided to drill Leyden No. 2 approximately 600 ft (182.88 m) south of the first location. This well was drilled in exactly the same manner as No. 1, except that it was deemed unnecessary to core drill the 200 ft (60.96 m) immediately above the cavern. Again the well bottomed out in an unmined area. At this point, it was decided to hydrofracture the two wells in an attempt to reach the cavern. The company was successful in fracturing out of No. 1, and after 3 days had pressure in the wellhead in excess of 8 psig (156.7 kPa), clearly indicating that there was a bubble of native air trapped in the structurally high portion of the mine.

The No. 2 well was less successful and has been abandoned. However, because there is some degree of communication between it and the cavern, it is now being used as a point of pressure observation.

To determine the possible degree of leakage, if any, before making the appreciable investments in line extensions for injection gas, portable compressors were used to pump about 10 MMcf (2.832×10^5 m³) of air into the virgin air bubble discovered at the bottom of well No. 1. This volume of injected air built the pressure up to 11 psig (177.4 kPa) which held constant for about 2 weeks. Although relatively insignificant in magnitude, the pressure results indicated that there was no sieve, and this evidence was sufficient to warrant the drilling of well No. 3 into the east workings. After this well was completed successfully, 4½ mi (7.24 km) of 18-in. (45.72-cm) line and 3½ mi (5.63 km) of 10-in. (25.4-cm) line were installed to wells No. 1 and No. 3 from existing transmission facilities. Injection of natural gas into well No. 3 was started immediately at a rate slightly higher than 1 MMcf/h (28.32×10^3 m³/h). This operation continued 8 hours per day for 1 week, at which time it was computed that the volume in storage was approximately 50 MMcf (1.416×10^6 m³) with a wellhead pressure of 93 lb (742.8 kPa). Near the end of the week, this pressure blew the water seal at the No. 4 air shaft, which is common to both workings, and as a result it was necessary to blow down or flare approximately 25 MMcf (7.079×10^5 m³) to relieve the pressure and reestablish the water seal at the shaft.

The evidence of gas tightness in the east workings was sufficiently encouraging so that well No. 4 was then drilled into the west workings. This well was successfully completed into the cavern; after the injection-withdrawal system was extended to it, operations were started to inject gas into the east and west workings simultaneously. As of December 31, 1959, there was in storage some 165 MMcf (4.672×10^6 m³) at an average pressure of approximately 50 psig (446.3 kPa).

This volume of gas in storage was of little or no value unless it could be delivered into the distribution system for consumer use. For this purpose, a 1 725-hp (1 286.3-kW) compressor station was constructed in 1959, complete with all necessary auxiliary equipment.

During the 1959–60 heating season, several trial runs were conducted during which relatively small amounts of gas were delivered into the distribution system. Such tests uncovered operational difficulties that had not been anticipated.

The objective of the program for summer and fall 1960 was to make the minimum storage volumes then developed operational for the 1960–61 heating season. The steps taken included drilling and connecting wells No. 5 and 6; installing drips and blow-off valves on low points in the injection-withdrawal system; and rebuilding wellheads so that freeze-ups could be remedied more easily and effectively. Experience gained over the previous 18 months indicated the possibility of considerably increasing the pressure on the cavern. Therefore, as a part of the 1960 summer program, it was decided to increase the cavern pressure to approximately 250 psig (1 825.3 kPa).

The injection program ended in mid-November 1960, at which time there was approximately 750 MMcf (2.124×10^7 m³) in storage at an average cavern pressure of 200 psig (1 480.5 kPa). During the remainder of the year, with the cavern shut in, pressure, water level observation, and leakage surveys continued with both conventional leak-detection equipment and a new infrared leak detector calibrated for methane gas. These observations showed both pressure and water levels remaining relatively constant, indicating that the structurally high portion of the cavern was gas tight. From these findings it was concluded that a minimum storage reservoir could be successfully completed and the company should proceed into the development phase of the entire mine.

Before the Leyden Mine could be fully developed, however, the hoisting and ventilating shafts had to be sealed. This was the most critical engineering phase of the work because these seals must be gas tight. Engineering of this problem had continued ever since the decision to proceed with the initial test phase of the project. After review of many proposals it was decided in favor of the method developed by the firm of Fenix and Scisson.

Figs. 19 and 20 show how this sealing was done. Fig. 19 is a detailed drawing of the No. 3 hoisting shaft. On completion of the cleaning out, all three compartments in the shaft were plugged at the bottom with concrete to a point 20 ft (6.10 m) above the bottom of the shaft. This plug was poured in two lifts of 12 ft (3.66 m) and 8 ft (2.44 m) each, with a 24-hour waiting period between pours. Next, the shaft was filled to the elevation of the final water level with common bank run gravel. At this point, the main concrete plug was poured in three equal lifts of approximately 6 ft (1.83 m) each. There was a 72-hour waiting period between pours. The actual configuration of the final design and placement of the main plug is in the form of a double-ended cork that keys the plug to the formation. The shaft was filled with 50 ft (15.24 m) of 3-in. (7.62-cm) to 6-in. (15.24-cm) igneous rock, 2 ft (0.61 m) of graded rock fill, 4 ft (1.22 m) of graded gravel fill, 4 ft of graded sand fill, and finally 20 ft of compacted clay. The shaft was filled with bank run gravel from the top of the compacted clay to the surface. The figure indicates a 6⅝-in. (16.83-cm) mud supply line that terminates in the rock fill between the top of the main concrete plug and the compacted clay plug. Eleven to 12-lb (4.98 kg to 5.44-kg) drilling mud was pumped into the

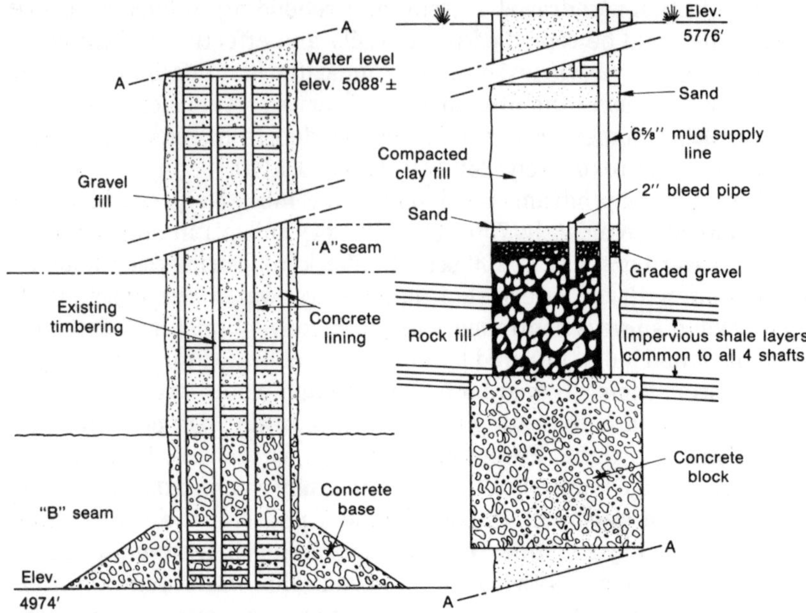

Figure 19. Detail of the Leyden Mine No. 3 hoisting shaft.

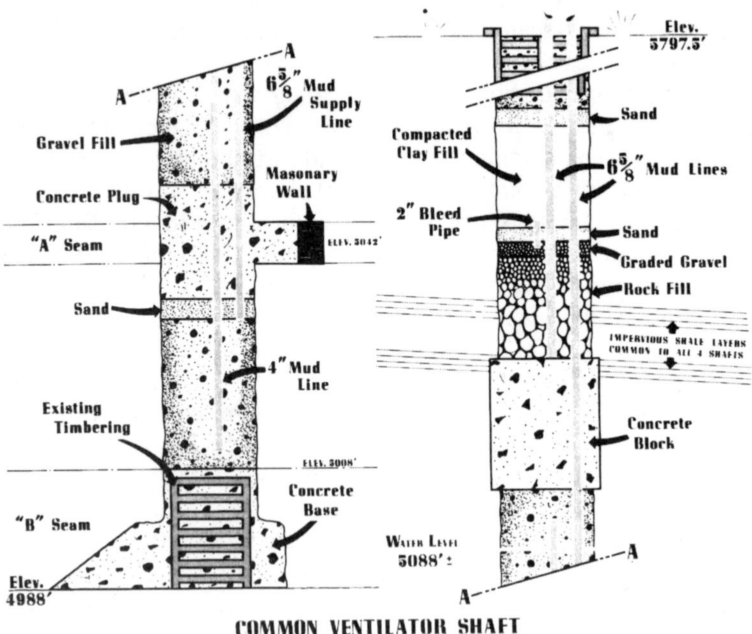

Figure 20. Detail of the Leyden Mine common ventilator shaft.

void spaces of the rock fill as the sealing agent, because the hydrostatic pressure exceeded the anticipated 300-psig (2 170.0-kPa) storage pressure and any leakage would be into the cavern. If mud loss continued, mud could be restored as it was dissipated from storage tanks located at each shaft; if mud loss were excessive, lost circulation material could be added to increase the effectiveness of the sealing agent.

Figure 20 is the detailed drawing of the common air shaft, where the plugging operation was the same, except that a secondary concrete plug was poured at the junction of the shaft and the A seam to effect a seal between the two mines so that they could be operated as independent storage reservoirs and add flexibility to overall operations.

Shafts 1 and 2 presented a somewhat different problem in that they were filled with debris and had to be cleaned out before sealing could begin. Afterwards, the sealing process was similar to that used in shaft No. 3. This construction began January 3, 1961, and was completed by early September 1961.

During the preliminary survey of shafts No. 3 and 4, an underwater, self-illuminated television camera was used to determine their condition below the water level.

The main concrete seals were installed in an impervious zone approximately 40 ft (12.19 m) above the A coal seam. On January 3, 1961, when a second consulting firm began work, this zone was under about 300 ft (91.44 m) of water, and it was necessary to withdraw the gas in the cavern to lower the water level. During a 3-month period beginning in mid-January 1961, the company withdrew 573 MMcf (1.623×10^7 m^3) of gas and pumped 103 MMgal (3.900×10^8 L) of water from the mine. To facilitate this process and meet the time schedule, two water wells and one additional gas well were drilled into the cavern.

After the shafts had been sealed, reinjection of gas under line pressure was begun in September, and as of November 30, 1961, 1.2 Bcf (3.398×10^7 m^3) of gas at approximately 210 psig (1 549.5 kPa) was in storage.

SAFETY DEVICES

During summer 1977, five retrievable-tubing safety valves were purchased and installed. These valves were installed in the better producing wells at an average depth of 300 ft (91.44 m). Eventually safety valves were installed in all 13 gas wells.

During the winter withdrawal period of 1979–80, while the safety valves were being routinely exercised, well No. 15 was found stuck in the open position. Because this well had a history of producing above-normal amounts of coal dust and grit, it was reasoned that particle accumulation was the culprit. Removal and inspection proved these suspicions correct. The program of safety valve installation was cancelled pending further study and operational evaluation.

PRESSURE REGULATION

If gas is purchased from the pipeline 4 mi (6.44 km) to the northeast, it is regulated by a remote-control regulator to the desired free-flow injection pressure. If gas is taken from the distribution system, it is necessary to compress for injection.

If the cavern is above 175 psig (1 308.1 kPa) during withdrawal, free flow from the cavern must be regulated to 175 psig for delivery into the distribution system. At lower pressures it is necessary to compress cavern gas for delivery at 175 psig.

MEASUREMENT

The measurement at Leyden consists of a single master meter station, composed of one 16-in. (40.64-cm) and two 20-in. (50.8-cm) orifice runs. This station is used during injection and withdrawal. During injection, gas is free flowed through one of the 20-in. runs if purchased from the pipeline company; if compressed from the distribution system, gas is flowed through the other 20-in. run. All three runs must be put in service to meet capacity during withdrawal. Because this facility is manned on a 24-hour basis, these runs are operated manually.

COMPRESSOR STATION

In 1962 five additional production wells were drilled; 2 000 hp (1 491.4 kW) of compression was added to the station; a second dehydrator, rated at 50 MMcf/d (1.410×10^6 m³/d), was installed. This increased the nominal withdrawal capacity to 100 MMcf/d (2.82×10^6 m³/d).

During summer 1963 another dehydrator, rated at 50 MMcf/d (1.410×10^6 m³/d), was installed. This one increased the capacity of the facility to 150 MMcf/d (4.230×10^6 m³/d). Also, pressure was increased to 250 psig (1 825.3 kPa) in the cavern for the first time, with the volume in storage at 2.1 Bcf (5.92×10^7 m³). Subsequent water pumping increased this volume to 2.6 Bcf (7.331×10^7 m³).

In March 1965 a late winter storm required a heavy withdrawal that tested the peak-day design capability of Leyden. On that day 150 MMcf (4.23×10^6 m³) of gas was successfully withdrawn.

High load growth continued in the Denver area during the early 1970's. New sources of peak-day gas supplies began to shrink. In August 1973 the company was notified by its major supplier that no additional peak-day supplies would be available for the 1974–75 heating season. The company was faced with the prospect of either curtailing additional firm growth on its gas system or resorting to its own means of supplying additional peak-day gas. In early 1974 it was decided to install a new dehydration tower to bring peak-day capacity from 160 MMcf/d (4.51×10^6 m³/d) to 185 MMcf/d (5.22×10^6 m³/d).

In the following seasons the company was faced with the same problem after two minor peak-day increases, although additional supplies were needed. After analyzing several peak-day storage alternatives such as supplemental natural gas, propane air, and LNG, management decided once more to expand the peak-day potential of Leyden. Doing so would require an incremental addition each year of the necessary plant and field facilities to sustain load growth.

A maximum deliverability of 230 MMcf/d (6.48×10^6 m³/d) was to be the ultimate design criterion based on new customer projections through 1980. This expansion required extensive rebuilding of the existing compressor plant, distribution system, wells, and gathering system.

The primary problem was to withdraw more peak-day gas from Leyden without increasing the storage volume from its physical limits of 2.6 Bcf (7.33×10^7 m³) at a top storage pressure of 250 psig (1 825.3 kPa).

To determine the total effect that an increase of 45 MMcf/d (1.27×10^6 m³/d) peak-day volume would have on the storage project, it was necessary to use a computer model to simulate the daily firm-gas requirements for each heating season of a 40-year weather history. An equation was used that represented projected daily firm-gas load at each mean temperature. This analysis indicated that during the coldest year, it would be necessary to sustain a continuous 5-day withdrawal of 230 MMcf/d (6.51×10^6 m³/d) to serve the firm-load requirements in the Denver and northern Colorado service area.

The study also indicated that, in all years other than the coldest-weather year, the number of days requiring maximum withdrawals of 230 MMcf/d (6.48×10^6 m³/d) would be minimal. In addition, a pressure-volume curve developed for Leyden indicated that a 5-day sustained withdrawal of 1 150 MMcf/d (3.24×10^7 m³/d) would reduce the cavern pressure to 100 psig (791.0 kPa).

As a result of summer withdrawal tests at Leyden, the company was able to establish a family of curves that represented the hourly and daily deliverability of the existing storage project at various cavern pressures. Because the existing system was capable of delivering only 230 MMcf/d (6.48×10^6 m³/d) down to a cavern pressure of 203 psig (1 501.2 kPa), it was necessary to design sufficient additional compressor horsepower to draw the cavern down to the required 100 psig (791.0 kPa) with a withdrawal rate of 230 MMcf/d.

It was necessary to plan a major reinforcement of the Denver distribution system to transport the additional large gas volumes on withdrawals. A direct benefit of reinforcing the distribution system for withdrawals proved to be the added ability to refill the mine to full storage pressure more rapidly between cold spells. As a result, the company can turn over relatively small storage volume several times during a heating season, and in effect the working volume is magnified.

Drawing the cavern down to 100 psig (791.0 kPa) left little additional pressure drop to be tolerated in the pipeline and well facilities between the cavern and the compressors. Because accurate underground survey records of the Leyden Mine were lacking, six of the existing injection-withdrawal wells were marginal producers. They had been drilled into pillars, rubble, or flooded areas.

Anticipating the same success ratio in drilling future wells, planners decided that the greatest value per dollar would be to reduce the pressure loss in the gathering system first. It was expected that the pressure loss in the gathering system could be reduced to approximately 10 psig (170.6 kPa) at maximum flow by looping existing large-diameter lines and replacing small lines.

Eliminating any additional pressure drop would then require the drilling of enough additional wells to reduce total pressure loss in the wells at maximum flow to approximately 30 psig (308.4 kPa). This pressure loss, plus the 10-psig (170.6-kPa) loss in the gathering system, resulted in a design suction pressure of only 60 psig (515.3 kPa) at the compressor station. Compressor discharge pressure of 200 psig (1 480.5 kPa) would be needed to overcome pressure losses in the plant and provide 175 psig (1 308.1 pKa) into the distribution system.

To provide 200 psig (1 480.5 kPa) would require a major addition of 10 500 hp (7 829.9 kW) of compression, complete with auxiliary equipment, to the existing 5 000 hp (3 728.5 kW), for a total of 15 500 hp (11 558.3 kW). This requirement was of great concern to the company because of the tremendous added weight of reciprocating compressor machinery and its supporting structures on a solid base that had already given problems at the existing plant.

The company had experienced unstable soil conditions under the existing plant structures due to a combination of expansive clays and a high water table. Therefore, in the evaluations of compressor equip-

ment, lightweight turbine engines and compressors were to be given
serious consideration.

After careful study of the above conditions, the company selected
three 3 500-hp (2 610.0-kW) Centaur turbine/compressor units manufactured by Solar Turbines International. The light weight and small
building requirements of this machinery enabled more efficient use of
the limited plant area and minimized the impact of additional weight
on the expansive soil.

Selection of three relatively small-horsepower compressor units,
rather than one or two large-horsepower reciprocating compressor units,
provided a minimum loss of horsepower during critical withdrawal
periods in the event of mechanical failure of one unit.

Limited construction time available between heating seasons also
favored the selection of solar compressor units because of the substantially less construction and equipment erection time required for small,
lightweight compressors. The greater flexibility of being able to add
only two turbine/compressor units in 1976 and one in 1977 provided
the opportunity to match horsepower requirements with load growth
on the system more closely.

In addition, turbine/compressors offered considerable savings in both
initial cost and subsequent maintenance expense. The high fuel use of
turbine/compressors compared with reciprocating compressors concerned the company at first, because of impending shortages of natural
gas.

However, a simulation of gas requirements showed that, even though
fuel use was high, the turbine/compressors would be used only during
a few "needle peak" days each year. The company would rely on its existing reciprocating compressor equipment to supply storage gas on all
other days requiring less than maximum delivery.

A visit to several turbine/compressor installations indicated a need
for reliable noise-suppressant facilities, both in the compressor building
and on the engine air-inlet system. Because residential encroachment
around the compressor plant was likely to continue, it was expected that
the public would not tolerate additional noise caused by the turbine/com-
pressor operation. Therefore, as the plant was being built during
1976–77, the walls of the compressor building were insulated and then
covered with perforated-steel sheathing to absorb and isolate the noise
generated inside. Next the engine air-inlet silencers, built and designed
for residential use, were mounted on the three turbine inlets. As a result,
the noise level at the property line was well below that required by code
during all turbine/compressor operations.

During the 1977–78 heating season, the Leyden storage project produced 196 MMcf/d (5.55×10^6 m³ /d). It was expected that as load growth

continued, additional well drilling would reduce the pressure loss at maximum flow until a design withdrawal of 230 MMcf/d (6.51×10^6 m³/d) for a 5-day sustained period is attained.

FIELD PIPING SYSTEM

The gathering system design includes wellhead separators, 10-in. (25.4-cm) and 12-in. (30.48-cm) laterals, 20-in. (50.8-cm) and 24-in. (60.96-cm) trunk lines, and a trunk line pigging system. There is a large stop tank collection vessel at the compressor station end of the gathering system, followed by three filter separators and the compressor suction headers.

GAS CONDITIONING PLANT

The discharge gas flows through gas after-coolers, filter separators, four dehydrator towers, an odorizer, and the meter station before entering the distribution system at 175 psig (1 308.1 kPa). The dehydrator has a single large reboiler and is designed to produce a gas stream with a maximum water vapor content of 7 lb/MMcf (11.13×10^{-5} kg/m³) of gas.

INJECTION

Injection occurs during September and October to bring the cavern up to its maximum working volume of 1 600 MMcf (4.53×10^7 m³) after its annual summer drawdown period. The maximum summer injection rate is 77 MMcf/d (2.12×10^6 m³) of gas purchased from a nearby pipeline at line pressure ranging from 325 psig (2 342.4 kPa) to 700 psig (4 927.9 kPa). Immediately downstream from the pipeline purchase point is a remote-control regulator controlled by the company's dispatching personnel. This regulator is used to limit injection to the maximum cavern pressure of 250 psig (1 825.3 kPa). This design eliminates the need for compression to inject during the annual refill period.

Injection also occurs in relation to heating-season loads and weather. If the mine has been used to peak shave and this period is followed by light loads, the mine can be refilled during the heating season at a maximum rate of 120 MMcf/d (3.40×10^6 m³/d). The objective is to purchase gas off-peak for refill on a daily basis. During the heating season, the cavern is refilled both with system gas (which has to be compressed) and with pipeline purchase gas.

WITHDRAWAL

Withdrawal occurs every spring, usually during late April and early May. The cavern pressure is reduced from 250 psig (1 825.3 kPa) to approximately 100 psig (791.0 kPa), a difference that represents about 1 150 MMcf (3.26×10^7 m³). The purpose of this drawdown period is to conduct performance tests on each well to facilitate routine maintenance to minimize leakage.

Like injection, withdrawal during the heating season depends on system requirements and weather. During this period, the Leyden Mine is used to avoid penalties when the existing daily contract supplies have been purchased. The maximum daily withdrawal rate of 185 MMcf/d (5.24×10^6 m³/d) can be maintained for 5 days, after which deliverability deteriorates quite rapidly.

RECYCLING

Fig. 21 illustrates the recycling of gas through Leyden. This chart is based on 1981–82 requirements, assuming that Denver experienced the severe weather of 1961–62. The maximum working volume at point 1 results from the annual injection period as previously described. At point 2, the daily contracts have been sold and it is necessary to use Leyden. This need continues for several days, causing the bottom of the trough to sink below the industrial rate line. At this point, some industrial customers must be curtailed to add to the Leyden supply. After the trough, demand decreases and Leyden is refilled over several days until it is at capacity toward the end of December. The remaining two troughs repeat the first, except point 3, which lies below the critical line. At this point, all industrials are curtailed to refill Leyden.

SHUT-INS

During the course of the heating season the cavern is maintained at a high cavern pressure to ensure deliverability. The shut-in periods associated with the heating season vary in length and depend on weather-related system requirements. While the pressure in the cavern is at 250 psig (1 825.3 kPa), it slowly drifts lower, sometimes as much as 1 psig/d. As this drift occurs, the mine is topped off to 250 psig as system requirements permit.

The longest shut-in period occurs each summer after the spring withdrawal and well-testing period. The cavern is shut in at 100 psig

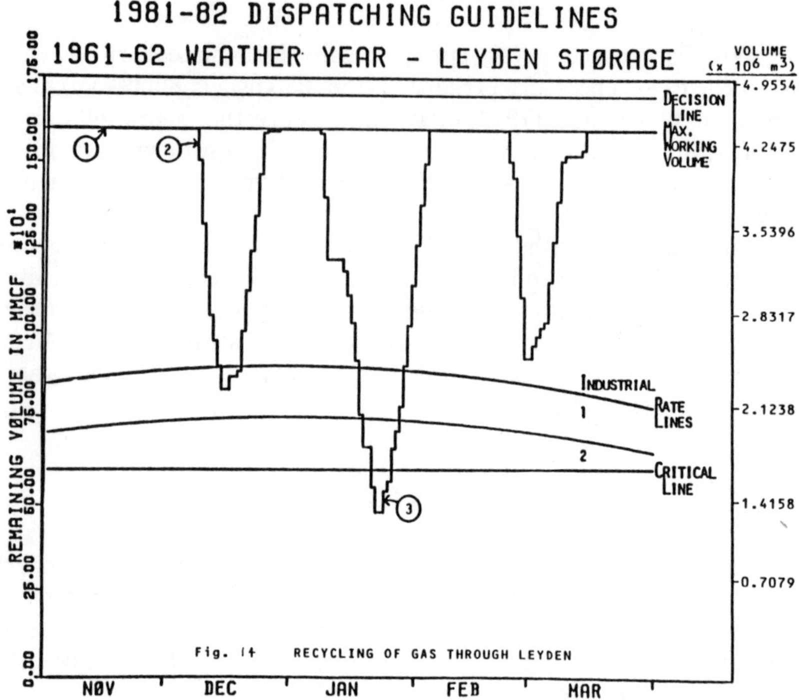

Figure 21. **Recycling of gas through Leyden.**

(791.0 kPa) for 120 to 150 days. During this period the cavern pressure slowly builds as gas migrates back into the cavern. It would not be uncommon to observe a pressure buildup from 100 psig (791.0 kPa) to 110 psig (860.0 kPa) at a decreasing rate over an 80-day period. The cavern pressure would stabilize near 113 psig (880.7 kPa) in about 120 days.

TESTING

The testing of each well takes place during the annual spring drawdown period. The purpose of this test is to ascertain performance data on each well for analysis. If the well's performance has decreased since previous tests, remedial work can be scheduled during the summer months. Performing this test requires shutting in all wells except the one being tested, which is then produced through the master meter station (wells are not separately metered) into the low-pressure distribution system.

WIRELINES

The use of wirelines other than for logging is limited to setting blanking plugs, which was necessary when safety valves were installed and serviced. Wirelines are also used when valve replacement on the wellhead is necessary.

WELL MAINTENANCE

To detect and locate leakage in the wells, a semiannual well logging program has been instituted at Leyden. A gamma-ray neutron log is run on each well at high pressure during the spring and at low pressure during the late summer. Once the logging is completed, the results are interpreted by a retained consultant, who recommends appropriate action.

REFERENCES

1. Van Horn, R., 1957, Bedrock Geology of the Golden Quadrangle, Colorado: *U.S. Geological Survey*, May GQ 103.

Chapter 4

DEPLETED CONDENSATE RESERVOIRS

Rapid River
ANR Storage Company
Michael J. Whims

INTRODUCTION

In this chapter, guidelines for developing depleted condensate reservoirs are drawn from the case study of the Rapid River 35 gas storage field in northern Michigan.

The FERC issued certification for developing the project in July 1979, and the project was in service in April 1980. The project involved converting two northern Michigan pinnacle-reef gas-condensate reservoirs to storage service, constructing a gathering system and compressor station at each location, and constructing 16 mi (25.7 km) of 36-in. (91.44-cm) pipeline to provide 38.3 Bcf (1.08×10^9 m³) of storage capacity for five interstate customers. This chapter details the engineering and operating practices used to develop one of the two fields. The two reservoirs, although different in storage capacity, were similar in their basic parameters. Identical design philosophy was used in the planning of the wells, gathering systems, and compressor stations for each location so that the two projects differed only in size and throughput capacity.

PRELIMINARY PLANNING

Gas has been stored in Niagaran reef reservoirs of western Ontario and southern Michigan since the 1940s. Although they were shallower and lower in original pressure, much valuable operating experience was obtained from these fields for developing engineering and operating philosophies.

GEOLOGY

The Rapid River 35 storage field is located in Kalkaska County, Michigan, within a producing area known as the Niagaran Pinnacle-Reef Belt of northern Michigan. The geology of the area has been effectively described in detail elsewhere. The 170-mi (274-km)-long reef belt is composed of approximately 500 oil-and-gas-condensate-producing pinnacle reefs, each of small areal extent [about 80 acres (32 ha) average] and high relief [up to 600 ft (183 m)]. They are composed of vuggy and permeable, dolomitized algal and coral reef material that was deposited at the northern edges of Michigan's Niagaran Sea some 400 million years ago. Beginning in Salina time, the reefs were covered with sediment and are now found at depths of 4 000 to 7 000 ft (1 200 to 2 100 m). Seismic investigation techniques have been used predominantly to delineate the individual reefs, and since first production in 1969, some 600 Bcf (1.69×10^{10} m³) gas and 120 MMbbl (1.91×10^{7} m³) of oil and condensate have been produced from the reef belt.

The Rapid River 35 reservoir was one of several nearly depleted gas-condensate reservoirs that were investigated for potential storage service. The field had been developed in 1973 and 1974 with two wells owned by major oil companies and a local Michigan firm. The reservoir was particularly attractive for several reasons:

- Its reserve capacity of approximately 15 Bcf (4.22×10^{8} m³) indicated it was a compact reservoir of small areal extent and high relief, lending itself to an economical well development program.
- Its location under flat, sandy terrain and near an existing pipeline right-of-way would reduce construction costs for the compressor station and pipeline.
- It was almost totally depleted, which would reduce the overall acquisition costs.

Copies of down-hole well logs and well completion records for the producing wells and surrounding dry holes were obtained to delineate the geology of the reservoir. The stratigraphic section shown in Fig. 22 was the result. It indicated that thick, evaporite rock units capped the reservoir. Values for net pay, porosity, and water saturation derived from the logs indicated that the reservoir should exhibit favorable storage characteristics. Structure and isopach maps were drawn from the available geological data to roughly delineate the areal extent of the

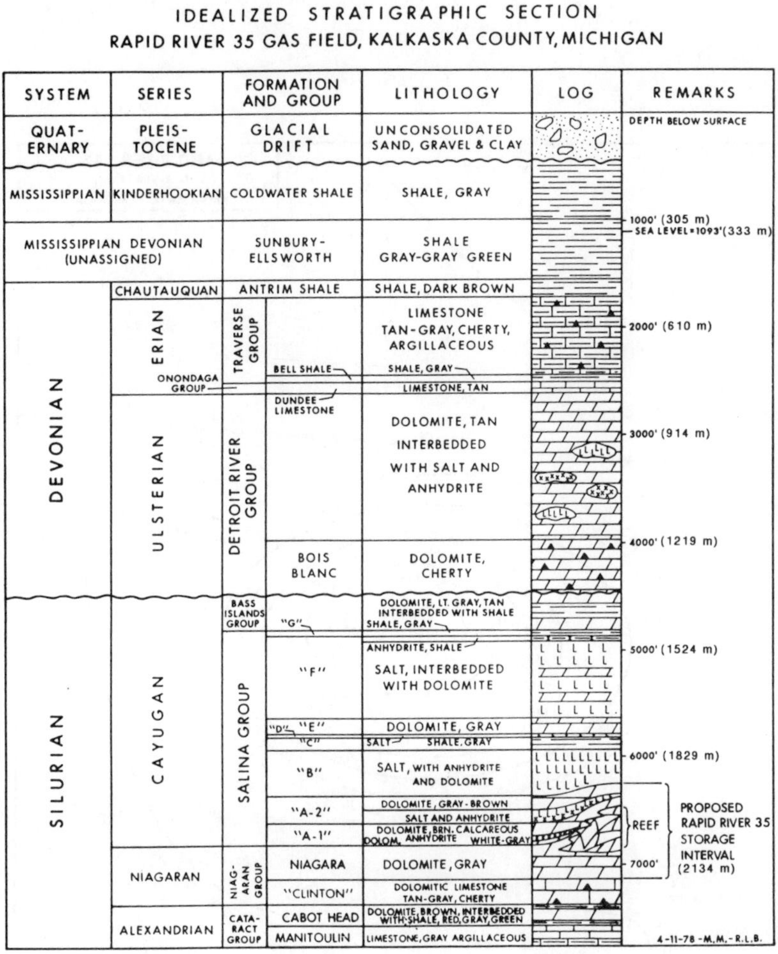

IDEALIZED STRATIGRAPHIC SECTION
RAPID RIVER 35 GAS FIELD, KALKASKA COUNTY, MICHIGAN

SYSTEM	SERIES	FORMATION AND GROUP	LITHOLOGY	LOG	REMARKS
QUAT-ERNARY	PLEIS-TOCENE	GLACIAL DRIFT	UNCONSOLIDATED SAND, GRAVEL & CLAY		DEPTH BELOW SURFACE
MISSISSIPPIAN	KINDERHOOKIAN	COLDWATER SHALE	SHALE, GRAY		1000' (305 m) — SEA LEVEL = 1093' (333 m)
MISSISSIPPIAN DEVONIAN (UNASSIGNED)		SUNBURY-ELLSWORTH	SHALE GRAY-GRAY GREEN		
DEVONIAN	ERIAN — CHAUTAUQUAN	TRAVERSE GROUP — ANTRIM SHALE	SHALE, DARK BROWN		
	ERIAN	TRAVERSE GROUP	LIMESTONE TAN-GRAY, CHERTY, ARGILLACEOUS		2000' (610 m)
	ONONDAGA GROUP	BELL SHALE	SHALE, GRAY		
			LIMESTONE, TAN		
	ULSTERIAN	DETROIT RIVER GROUP — DUNDEE LIMESTONE	DOLOMITE, TAN INTERBEDDED WITH SALT AND ANHYDRITE		3000' (914 m)
	ULSTERIAN	DETROIT RIVER GROUP			4000' (1219 m)
		BOIS BLANC	DOLOMITE, CHERTY		
SILURIAN	CAYUGAN	BASS ISLANDS GROUP — "G"	DOLOMITE, LT. GRAY, TAN INTERBEDDED WITH SHALE SHALE, GRAY		
		SALINA GROUP — "F"	ANHYDRITE, SHALE SALT, INTERBEDDED WITH DOLOMITE		5000' (1524 m)
		"D" "E"	DOLOMITE, GRAY		
		"C"	SALT SHALE, GRAY		
		"B"	SALT, WITH ANHYDRITE AND DOLOMITE		6000' (1829 m)
		"A-2"	DOLOMITE, GRAY-BROWN SALT AND ANHYDRITE		PROPOSED RAPID RIVER 35 STORAGE INTERVAL (2134 m) — REEF
		"A-1"	DOLOMITE, BRN. CALCAREOUS DOLOM. ANHYDRITE WHITE-GRAY		
	NIAGARAN	NIAGARAN GROUP — NIAGARA	DOLOMITE, GRAY		7000'
		"CLINTON"	DOLOMITIC LIMESTONE TAN-GRAY, CHERTY		
	ALEXANDRIAN	CATARACT GROUP — CABOT HEAD	DOLOMITE, BROWN, INTERBEDDED WITH SHALE, RED, GRAY, GREEN		
		MANITOULIN	LIMESTONE, GRAY ARGILLACEOUS		4-11-78 -M.M.- R.L.B.

Figure 22. **Idealized stratigraphic section of Rapid River 35 gas field.**

reservoir. This preliminary work indicated that the storage reservoir was a pinnacle reef with 460 ft (140 m) of gross thickness, overlain by a "super caprock" composed of salt and anhydrite 35 ft (10.7 m) thick at its minimum. The reservoir averaged 233 ft (71 m) of net pay, at 9 percent porosity with very low water saturation, and had an areal extent of 160 acres (65 ha).

After all other alternatives were weighed against the design conditions for the entire project, the decision was made to begin negotiations to acquire the storage field. Producer seismic information was obtained

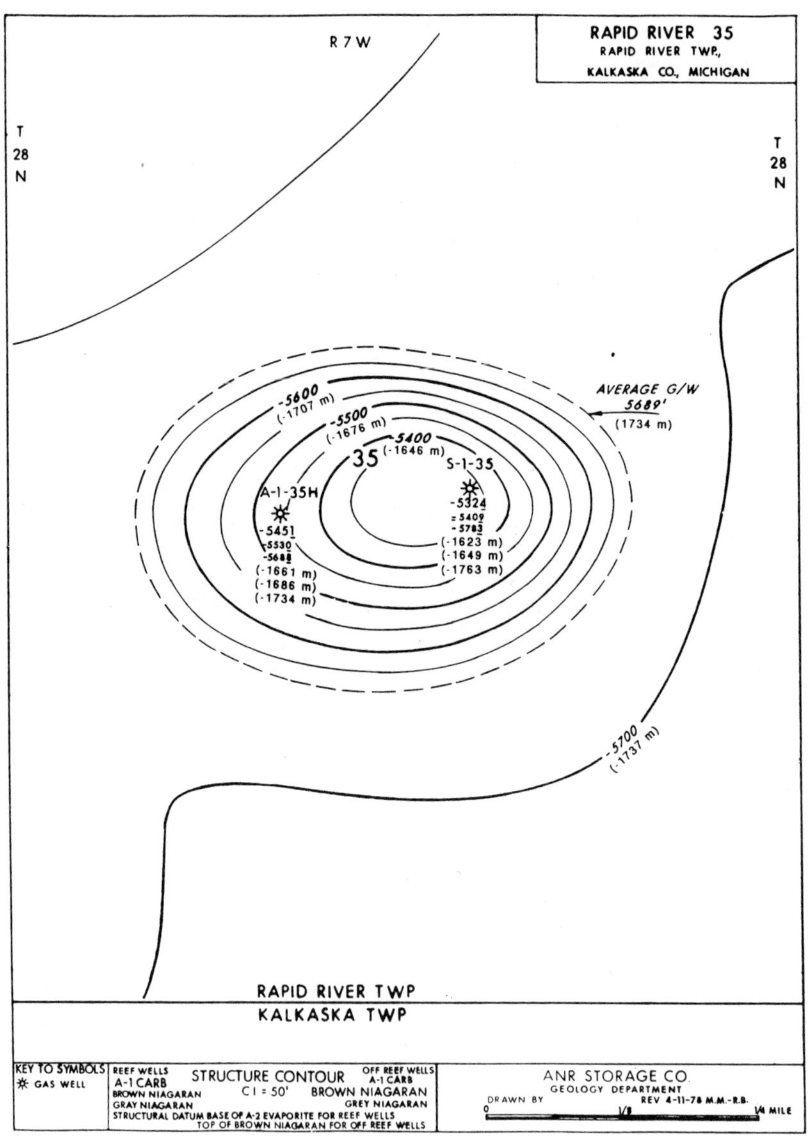

Figure 23. **Final structure map of Rapid River 35 gas storage field.**

for the field as well as the surrounding area. The seismic information was used to delineate the shape of the field and to help verify that the reservoir was a distinct "pressure vessel" not associated with another reservoir. A final structure map was prepared (see Fig. 23). At this point, everything that could be interpreted geologically from the available data and seismic had been done. Further refinement using deliverability data and production information before choosing well bottom-hole locations is discussed later in the "Reservoir System" section.

LAND AND RIGHT-OF-WAY

To ensure that no unusual land acquisition problems would be encountered, land records were reviewed before the decision was made to proceed with the project. It was decided that first efforts would be directed toward acquiring the working interests attributable to each producing well in the field as a basis for securing royalty and other interests.

Michigan law specifies the size for drilling units through the supervisor of wells, Michigan Department of Natural Resources. Each of the two original producing wells in the storage field was drilled on a 160-acre (65-ha) drilling unit (see Fig. 24).

Production proceeds from the two wells had therefore been historically distributed among the working interests ($\frac{7}{8}$) and the royalty interests ($\frac{1}{8}$) within the 320-acre (130-ha) area. Michigan law also provided for the further proration of gas production by the Michigan Public Service Commission within a field of two or more wells, based on a formula that adjusted for the drainage area of each producing well. This two-well field was produced under such a proration formula, which further defined the distribution of production proceeds among working and royalty interests. On the basis of these adjusted distribution decimals, as well as recorded production and reserve information, dollar values for remaining reserves in place were determined and agreed upon between the producers (working interests) and the storage company. The existing wells and wellhead equipment were also purchased from the working interests at this time.

The mineral owner (in this case the State of Michigan) was then contacted, and its royalty interest in the remaining reserves in the producing units was purchased based on the results of the working interest acquisitions. Overriding royalty interest was acquired in like manner, so that the totality of all decimal interests acquired would equal one (1.0). At this point, the storage company had acquired the original producers' position within the terms and conditions of existing gas purchase contracts with local gas purchasers.

Private mineral interest for properties lying outside the 320 acres

(130 ha) of producing units but within the storage acquisition boundary
(shown in Fig. 24) were then acquired on a dollars-per-acre basis. Sim-
ilarly, private storage rights for all property inside the storage acquisition boundary were acquired on a dollars-per-acre basis.

Surface easements for the construction of the gathering systems
and compressor station were then acquired from the state by lease rental
agreement. These easements covered an area of 43.35 acres (17.54 ha)
for the storage facilities. Pipeline right-of-way easements were obtained
from both private and state landowners in the form of rentals for the
16 mi (25.7 km) of 36-in. (91.44-cm) pipeline that would be installed.

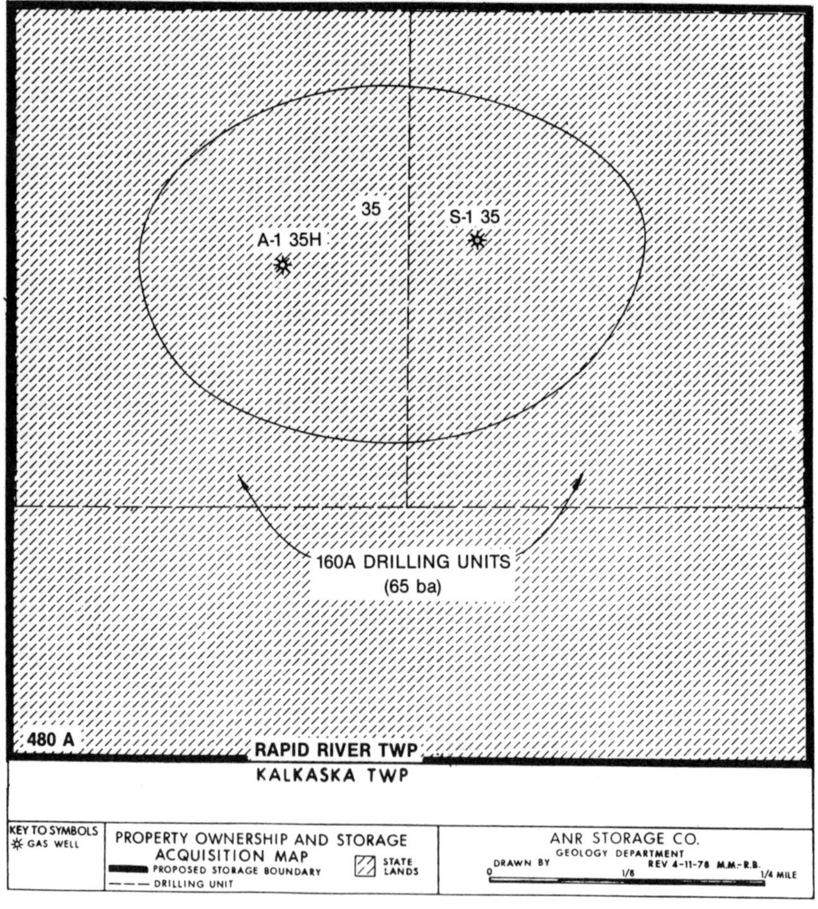

Figure 24. **Property ownership and storage acquisition map for Rapid
River 35 storage field.**

REGULATORY REQUIREMENTS

Because the project involved interstate storage of natural gas, it was necessary to apply to the FERC for the authority to provide storage service. The application was filed in July 1978, and companion applications by various other companies providing transportation service were filed shortly afterward. Temporary certification for the construction of storage facilities was received in February 1979, and the final certification of the entire project was received in July 1979.

As required by the Clean Air Act for gas compressor installations that have the potential to emit more than 250 tons per year of any regulated pollutant, an application for the storage project was filed with the EPA in June 1978. Certification for the compressor installations was received in September 1979.

To secure the necessary flexibility to drill the new storage wells where needed, the 160-acre (65-ha) well spacing order in force for the storage field had to be abrogated. A petition was filed before the supervisor of wells, Michigan Department of Natural Resources, in April 1979. The petition requested that the supervisor of wells abrogate the existing well spacing restrictions and dedicate the Salina-Niagaran formation to storage within the confines of the storage acquisition outline. A public hearing was held in May 1979 before the Michigan Oil and Gas Advisory Board, to take testimony from the storage company and other interested parties about the need for the storage project, the fairness to all parties concerned, and the technical feasibility of the project. The board subsequently voted to make a favorable recommendation to the supervisor, who issued an order in June 1979 abrogating well spacing within the storage acquisition outline as it related to storage wells drilled into the storage field only. All other wells must continue to be spaced under previous standing orders, subject to intervention and review by the storage company. Thus the interests of future exploration as well as gas storage were served.

Permits for the drilling of 13 storage wells were applied for in April 1979 through the supervisor of wells. The permit applications contained the required well design detail, certified subsurface directional programs, a registered land survey of each location, blowout prevention programs and equipment, and an environmental impact assessment for each well site. Drilling permits were issued in July 1979. A rework permit was required for only one of the existing wells, because changes had been made to its casing (perforations were added).

Other permits, such as building permits, water well permits, and sewer permits, were obtained from local government bodies.

RESERVOIR SYSTEM

Before the decision was made to develop the storage reservoir for storage service, many reservoir engineering studies were made in conjunction with the previously described geological studies to determine the reservoir's suitability for our storage needs. Gas production records obtained from the Michigan Public Service Commission and the Michigan Department of Natural Resources were cross-checked and verified. Condensate production records and bottom-hole pressure records were obtained from the Michigan Department of Natural Resources. Two-phase deviation factors (z factors) were roughly determined from actual data and assumed PVT relationships. A curve for bottom-hole pressure (BHP/z) versus cumulative hydrocarbon production was constructed from the data to determine original gas in place. This curve was then used to verify the size of the field as drawn on the preliminary structure map. The straight-line fit of the data points on the curve also verified that the reservoir had a pressure-depletion drive and that there was no apparent pressure communication between the storage reservoir and any other reservoirs. The volume of condensate production shown on the state records indicated that the reservoir was indeed a gas-condensate reservoir.

After negotiations with the working interests had begun, the producers' records of gas, condensate, and water production were acquired, as well as bottom-hole pressure surveys, PVT studies, and gas and liquid analyses. (Deliverability tests and drilling records also acquired at this time are discussed in the "Deliverability" section.) The production data and bottom-hole pressure surveys were cross-checked with the data previously received and verified for accuracy. Two-phase deviation factors (z factors) were obtained from the PVT studies, and were substituted for the calculated z factors previously used. A new curve of BHP/z versus cumulative hydrocarbon was constructed, which indicated gas in place amounting to 15.2 Bcf (4.28×10^8 m³). Table 1 lists the pertinent production and reserve data for the Rapid River 35 Field.

Producers' production records were apparently incomplete on the subject of brine production for many Niagaran reef reservoirs. Although the records indicated little or no brine production, it was determined through conversations with the producers' field operators and the gas purchasers' field operators that salt buildup in the field piping and well tubing had been an occasional problem. Down-hole well logs verified the existence of a small brine leg in the storage reef. Brine analyses from nearby reefs showed this brine to be a sodium chloride-calcium chloride mix with a density of about 10.6 lb/gal (4 611 kPa/m). As in most of these reefs, the producers had perforated the storage field wells at or near the apparent gas water contact in hopes of maximizing the

TABLE 1
Production and Reserves Data for Rapid River 35 Field

Original gas in place, 14.73 psia (101.56 kPa)	15.2 Bcf (4.28×10^8 m³)
Cumulative gas production, 14.73 psia (101.56 kPa)	13.7 Bcf (3.86×10^8 m³)
Cumulative condensate production	255 000 STB (40 542 m³)
Initial condensate yield	46 STB/MMcf (2.6×10^{-4} m³/m³)
Discovery reservoir pressure	3 485 psia (24 028 kPa)
Reservoir temperature	118 °F (48 °C)
Datum reservoir depth	6 704 ft. (2 043 m)

production of liquid hydrocarbons. Because of the presence of a brine leg, the decision was made to complete the storage wells approximately 100 ft (30.5 m) above the apparent gas-water contact and to install brine separation and storage facilities at the surface.

Determining residual water saturation of the reservoir pore spaces turned out to be difficult. Down-hole well logs for the existing wells were digitized and analyzed on an in-house computer program and seemed to indicate water saturation values ranging from 0 percent to 20 percent throughout the entire net pay section. There appeared to be some correlation between low-porosity zones and zones of high water saturation, suggesting drilling fluid invasion, but the correlation was not complete. It was known, however, that very little if any water enrichment of the storage gas had occurred in similar reef fields in southern Michigan generally.

To determine whether expensive gas dehydration facilities were needed at Rapid River 35, a more thorough investigation was begun. The producers' and gas purchaser's records were searched for water dew point values. To supplement these data, approximately 50 gas fields in various stages of depletion were sampled and a general correlation was developed, which indicated that the gas was indeed fully water saturated at discovery pressure conditions but became more undersaturated as depletion progressed. The initial decision was made to omit gas dehydration facilities from the surface equipment design. This topic is discussed further in the Surface Facilities section, page 112.

GAS INVENTORY

Determination of working storage, base gas, top pressure gradient, and base pressure level was structured on experience with similar reef storage fields in southern Michigan as well as on available data on the subject storage field. Previous caprock studies of the anhydrite caprocks of southern Michigan Niagaran reefs had labeled them "super caprocks," essentially impermeable to gas flow at threshold displacement pressures up to 2 381 psig (16 416 kPa) under laboratory conditions. Because the caprock of northern Michigan reefs is also anhydrite, the decision was made to design the storage field for the same top reservoir pressure gradient currently used for southern Michigan reef storage fields: 0.72 psi/ft (16.29 kPa/m) of depth. The result was a top storage pressure of 3 960-psig (27 303-kPa) wellhead pressure, or top caprock threshold displacement pressure of 1 160 psig (7 998 kPa). The use of a 0.72 psi/ft (16.29-kPa/m) gradient was to be contingent on the findings of the storage well-drilling program, however. A base wellhead pressure of 500 psig (3 447 kPa) was determined through a cost-benefit study that investigated the most cost-effective combination of wells, compressor horsepower, and base gas. Table 2 presents the resulting gas inventory data for the storage field. These data assume that the injection and withdrawal gas is "dry," i.e., contains no hydrocarbon liquid.

A major factor that had to be considered in developing the storage reservoir for storage service was the assumed presence of retrograde condensate in the formation. The production data in Table 1 show that the field had already produced 255 000 STB (40 542 m³) of condensate, with a historical average yield of 18.7 bbl/MMcf gas (1.05×10^{-4} m³/m³). Several questions had to be answered:

- Is Rapid River 35 truly a gas-condensate reservoir, exhibiting the corresponding retrograde condensate phenomenon?
- If so, what recoveries should be expected?
- How will the retrograde condensate in the reservoir affect storage service?
- How long will the retrograde condensate effects be felt?

Rapid River 35 was determined to be a true gas-condensate field that had recovered less than half of the original condensate in place. The process of injecting lean gas to pressures over the dew point would be expected to revaporize the retrograde condensate in the reservoir

TABLE 2
Gas Inventory Data for Rapid River 35 Field

Working storage capacity, 14.73 psia (101.56 kPa)	14.7 Bcf (4.14×10^8 m³)
Base gas	2.3 Bcf (6.48×10^7 m³)
Top storage pressure, wellhead	3 960 psig (27 303 kPa)
Top storage pressure, bottom hole	4 634 psig (31 950 kPa)
Base pressure, wellhead	500 psig (3 447 kPa)
Depth to caprock	6 436 ft (1 962 m)
Maximum storage pressure gradient	.72 psi/ft (16.29 kPa/m)

and produce a certain amount of condensate each withdrawal season for a short but unknown period. For purposes of surface facility design, a maximum figure of 2 500 bbl/d (397 m³/d) was used, based on the sweep efficiency calculations of the report. Injection-withdrawal well usage (sweep patterns) were found to change dramatically the theoretical recovery of condensate. The actual location of the retrograde condensate in the pore spaces, whether scattered throughout the rock or collected on bottom, could greatly change the recovery estimates. The heating value of the withdrawal gas should be expected to be high for a number of seasons because of liquid enrichment, requiring a means of reducing the heating value of the gas within the surface equipment design.

Many of the report's findings were experimental and involved some critical assumptions and new techniques. To verify the liquid-related findings and theories, as well as provide acceptable monitoring of the gas inventory, a comprehensive monitoring system was designed for the field. A system of gas and liquid sample points was designed to keep track of injected and withdrawn hydrocarbons by component as well as by phase. Verification of gas inventory and identification of losses is more difficult because of the anticipated pressure-content effects of the retrograde condensate. Furthermore, these effects can be different each succeeding season as liquid depletion continues. With vigilance on the part of the engineers and field operators, the potential of retrograde condensate did not appear to be a problem.

DELIVERABILITY

Original back-pressure tests of the two existing wells in the storage field were obtained from the producers. After verification that the tests were reasonable and accurate, their values were averaged to construct an "average well," bottom-hole deliverability curve. Based on this curve, wellhead deliverability curves were constructed for various casing and tubing sizes. After analysis for cost-effectiveness and engineering feasibility, the decision was made to use 7-in. (17.8-cm) casing as flow tubulars for the wells.

The storage service contracts specified the required deliveries by month during withdrawal. Because the "worst case" withdrawal condition for the wells would probably be in late March, the number of wells was based on late March withdrawal conditions. An in-house computer program for well interference was used to simulate the effect of well interference at the worst-case condition and at a given well-placement pattern. Calculations indicated that 13 new wells would be needed to provide the required storage service.

WELL DESIGN

The wells were designed around the need for 7-in. (17.8-cm) flow tubulars to the reservoir at approximately 6 500 ft (1 981 m). Michigan law also required the wells to incorporate an intermediate casing string to seal off potential loss zones above approximately 5 000 ft (1 524 m). It was believed imperative for storage service that all casing strings be cemented to surface whenever possible to minimize corrosion and undue casing flexure during injection and withdrawal. The availability of certain standard casing sizes in Michigan was also considered in sizing the casing strings. In all cases, design factors for burst, collapse, and tension were included in the design of the wells. The storage well design included 20-in. (50.8-cm) conductor pipe to 100 ft (30.5 m), 13⅜-in. (34-cm) 8-rd surface casing to 700 ft (213 m), 9⅝-in. (24.4-cm) 8-rd intermediate casing to 5 000 ft (1 524 m), 7-in. (17.8-cm) buttress production casing to 6 500 ft (1 981 m). All casings were cemented to the surface if possible.

DRILLING

The storage field was located on state forest land, like most of the northern Michigan fields. Environmental considerations dictated minimizing the amount of forest clearing done to drill the storage wells. It was decided that the storage wells would be drilled directionally from centralized pads. For safety reasons, the surface well locations were spaced along a straight line, 150 ft (45.7 m) apart. Two pads were designed, running roughly north and south, one containing six new wells and one existing well, and the other containing seven new wells and the other existing well. In this way, all areas of the reservoir could be covered by reasonable hole deviations.

One of the major design considerations for the drilling program was the probable loss of circulation to be encountered in penetrating the 6 500-ft (1 981-m)-deep reef, which contained a bottom-hole pressure of only 360 psig (2 482 kPa) at the time. Vugular porosity seen on the original well logs indicated isolated permeability in the multidarcy range. This situation would also severely reduce the probability of obtaining full cement returns to the surface. It was therefore decided to drill only 5 to 10 ft (1.5 to 3 m) below the anhydrite caprock into the tight upper part of the reef, set the 7-in. (17.8-cm) casing, and cement it to the surface. The well would then be completed open-hole with a smaller completion rig. The wells were drilled with an Ideco Hydrair H44CD rig, rated for 7 500 ft (2 286 m). A gelled water-drilling system was used for the entire hole. Through the salt sections to the 7-in. (17.8-cm) casing point the drilling fluid was saturated with salt to reduce hole washout. Down-hole motors were used where needed to establish hole direction, but all other directional drilling was done with standard rock bits and drilling assemblies. Cement bond logs indicated good to excellent cement jobs on the 7-in. casing. In all cases, the bottom few hundred feet of pipe, which was centralized, exhibited excellent cement bond. No special preparations were made to the casing other than specifying no mill varnish from the mill and then allowing the pipe to weather before being run in the well.

CORING AND LOGGING

Because of the probability of severe circulation loss while attempting to core the reef, no coring was done. Drill-cutting samples from two of the wells were analyzed through the reef section for the presence of condensate by means of a newly developed pyrochromatigraphic process. The results indicated that the retrograde condensate was distributed throughout the reef.

The logging program included a full suite of electrical, acoustical, and radiation logs on the entire hole for three of the wells. The remainder were logged from the intermediate casing string down through the reef only.

COMPLETION

A smaller double rig rated for 215 000-lb (97 522-kg) capacity was moved on the hole for open-hole drill-in and completion. After drilling out the cement float collar and cement to the guide shoe, the casing was hydrotested to bottom-hole pressure of 4 600 psig (31 716 kPa). A second gyro survey was run to verify the results of the single-shot surveys used during drilling. The open-hole section was then drilled with fresh water and a down-hole motor. The use of a down-hole motor eliminated the damage to the inside of the 7-in. (17.8-cm) casing that would have been caused by conventional rotating techniques. In most cases, drilling returns were not complete. After the reef had been drilled to a depth of 100 ft (30.5 m) above the apparent gas-water contact, the drill string was pulled out while the hole was kept full. Acid-soluble, lost circulation material used on the first few wells proved ineffectual. The hole was then logged and blown dry.

A permanent-retrievable packer was then set by electric line approximately 100 ft (30.5 m) above the 7-in. (17.8-cm) casing shoe. Below the packer, a 2½-in. (6.35-cm) landing nipple had been installed, and into this a wireline plug was then placed. The dry casing was then filled with nitrogen and the wellbore tested to 4 000-psig (27 579-kPa) surface pressure to simulate maximum storage conditions and then blown down. An upper permanent-retrievable packer was then installed approximately 100 ft (30.5 m) below ground surface. A 2½-in. landing nipple was installed below this packer. Into this upper packer was installed a 2½-in. tubing-retrievable, surface-controlled, subsurface safety valve, which was run in on 2⅞-in. (7.30-cm) tubing and landed into the tubing head at surface. The safety valve was hydraulically controlled at the surface through a ¼-in. (0.64-cm) control line down to the safety valve. The 2⅞-in. tubing immediately above the safety valve had been perforated to allow the gas to reenter the 7-in. casing as it flowed from the safety valve to the surface, and vice versa. The final wellbore configuration can be seen in Fig. 25. A 5 000-psig (34 474-kPa) working-pressure wellhead was then installed at the surface. After the safety valve and upper assembly were tested, the wireline plug was removed

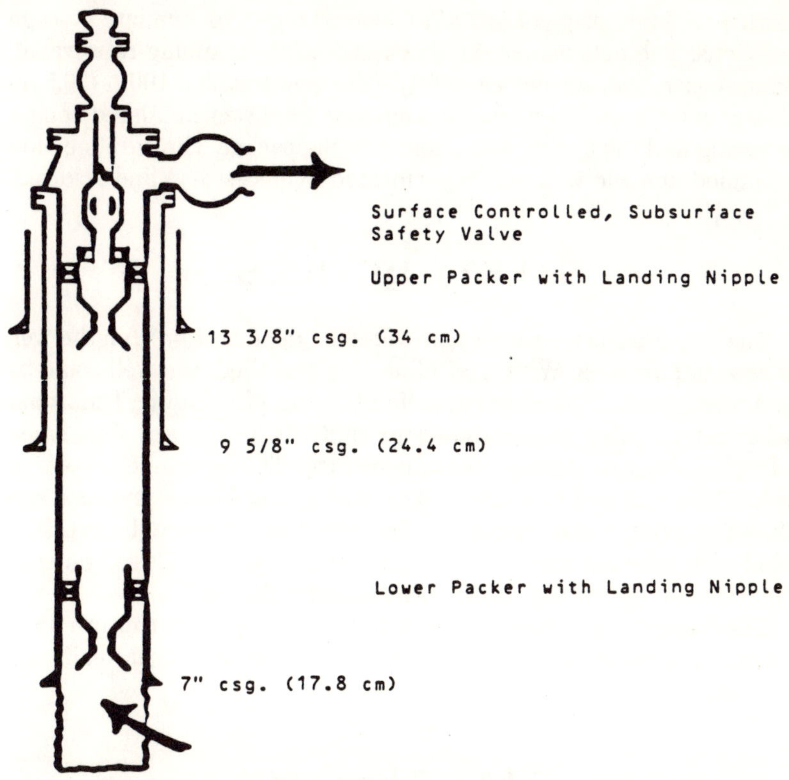

Figure 25. **Typical storage wellbore configuration of Rapid River 35 field, Kalkaska County, Michigan.**

and the well was acidized with 10 000 gal. (37.85 m³) of 28-percent hydrochloric acid in an 80-quality foam. The well was then blown dry and, in some cases, flow tested with a critical flow prover.

The two existing wells were evaluated for use as injection-withdrawal wells. The old 2⅞-in. (7.30-cm) tubing was pulled and discarded. The casing was checked for integrity with a wireline corrosion-monitoring log; the original cement job was analyzed by means of a cement bond log. A permanent-retrievable packer with lower landing nipple was installed a few feet above the existing perforations. A wireline plug was installed in the landing nipple and the hole was refilled with water. The old casing and tubing spools at the surface were then replaced with 5 000-psi (34 474-kPa) working pressure (WP) equipment. New 2⅞-in. (7.30-cm) ABC modified tubing was run back in the hole to facilitate blowing the hole dry with nitrogen. The well was then tested to wellhead pressure of 4 000 psig (27 579 kPa) with nitrogen to simulate storage conditions. The hole was blown down and a 2⅞-in. tubing-retrievable, surface-controlled, subsurface safety valve was installed 100 ft (30.5 m) below ground level. Then inhibited annulus fluid was circulated behind the tubing and the tubing was stung into the packer. The wireline plug was pulled and the well was reperforated (if necessary) and acidized.

SAFETY DEVICES

The lower packer assembly provided a "false bottom" for the well for rework purposes. With a wireline plug installed, the well could be filled with water, if necessary, without losing circulation. The upper packer-safety valve was designed to shut off all upward flow when hydraulic pressure is bled from the surface. The hydraulic system at the surface was tied to a fire-sensitive plug that bleeds pressure and closes the safety valve in case of fire. The loss of hydraulic pressure would also activate and close the gathering system lateral valve to eliminate the possibility of back-flow toward the well.

The casing annuli were fitted with relief valves and rupture disks to ensure against dangerous buildup of annulus pressure that might rupture outer casing strings.

WELL TESTING

Well tests performed during the drilling program to check the deliverability calculations indicated that the actual field deliverability had been underestimated. The average well CAOF potential was found to be more than 600 MMcf/d (1.69×10^7/d m³/d). The result was a reduction in the number of new wells required from 13 to only 7.

SURFACE FACILITIES

Because limited time was available for the engineering design and construction of this project, because the members of the design group were unfamiliar with the high operating pressures, and because of the uncertainties involved with developing a gas-condensate reservoir for gas storage, a project team concept was used for project development. The project team included members with backgrounds in compressor engineering, pipeline engineering, electrical and controls engineering, reservoir engineering, construction, purchasing, drilling, land acquisition, law, environmental, and field operations. The team met weekly in Detroit and made dozens of trips to various manufacturers and other high-pressure storage operations, and some of the team members spent months with outside engineering firms and manufacturers in order to design the facilities quickly, efficiently, and safely. Because the design conditions often represented a "whole new ball game" to the project team, objective and even novel engineering was incorporated in the design. And because some of the team members were veteran storage field operators, the facilities were designed with operational feasibility in mind.

Beginning at the wellheads on withdrawal, a 4 000-psig (27 579-kPa) gathering system was designed with common injection and withdrawal piping. Six-in. (15.2-cm) laterals were used to tie each well to common 12-in. (30.5-cm) headers. Each lateral contained an electrically operated valve at a point 150 ft (45.7 m) from each wellhead for controlling well flow manually or remotely. A 6-in. (15.2-cm) gate valve was located at each wellhead for backup. An orifice fitting and numerous pressure, temperature, and sampling taps were also put in each lateral. The two 12-in. (30.5-cm) headers (one from each drilling pad) converged at the compressor station fence into a common slug catcher. This 4 000-psig (27 579-kPa) WP slug catcher was designed to handle high-velocity liquid that tends to build up in the gathering system or wellbores. Liquid is sent directly to the low-pressure liquid-handling facilities after pressure reduction.

Gas exits the slug catcher, is heated through two indirect fired heaters (if necessary), and passes into the metering facilities. Metering is done through one 12-in. (30.5-cm), 4 000-psig (27 579-kPa) WP Senior orifice fitting. The gas then enters the pressure regulation facilities, composed of two 8-in. (20.3-cm) noise-abating, two-phase pressure reduction valves. These valves reduce withdrawal gas pressure from a maximum inlet pressure of 4 000 psig to a normal outlet pressure of 900 psig (6 205 kPa). The outlet gas temperature can be controlled to approximately 35 °F (1.7 °C) through gas heating and Joule-Thompson cooling until inlet pressure declines (as field pressure declines) to approximately 1 800 psig (12 410 kPa). At that point, further cooling can be accomplished by gas recirculation.

The cooling accomplished by gas pressure regulation is designed to remove condensable liquids and control gas heating value. The multiphase stream then flows into an 8-ft by 32-ft (2.4-m by 9.8-m), 1 050-psig (7 239-kPa), WP, horizontal, three-phase separator/mist extractor. Water is dumped directly to stock tanks. Condensate leaves the three-phase separator, goes to a metering facility and exits the storage site via liquid pipeline.

Gas leaving the three-phase separator can follow any of three paths:

- If the pressure is sufficient to allow free flow into the gas pipeline, the gas bypasses all downstream facilities and directly enters the gas pipeline.
- If pressure is not sufficient to allow free flow into the gas pipeline (low field pressure conditions), the gas passes through a 5-ft by 15-ft (1.5-m by 4.6-m) filter separator and is compressed by two 3 750-Hp Ingersoll-Rand 410-KVR compressor units, is cooled through aerial coolers, and then enters the gas pipeline.

Gas entering the gas pipeline travels from the Rapid River 35 field approximately 16 mi (25.7 km) through a 36-in. (91.44-cm), 1 050-psig (7 239-kPa) WP pipeline to a metering station. The 16-mi pipeline contains pigging facilities at each end and is also equipped with several pipeline drips at strategic low points as it traverses the hilly glacial terrain of Kalkaska County. The gas is metered, and gas properties are measured for custody transfer on a heating-value basis.

During injection, the reverse path is followed through the 36-in. (91.44-cm) pipeline to the compressor station. The gas passes through the scrubber and the filter separator and is compressed by the same compressor units up to 4 000 psig (27 579 kPa). The high-pressure gas is then cooled through the aerial coolers, is metered through the same high-pressure metering facilities, and passes through the same gathering system to the wells.

OPERATIONS

This section discusses first-year injection and withdrawal operations.

INJECTION

Gas was injected to a top wellhead pressure of 3 914 psig (26 986 kPa) at rates up to 208 MMcf/d (5.86×10^6 m³/d). Total content in storage at the end of injection was 17.4 Bcf (4.9×10^8 m³), which was slightly higher than the anticipated 17.0 Bcf (4.8×10^8 m³).

WITHDRAWAL

Maximum gas withdrawal rate was 301 MMcf/d (8.48×10^6 m³/d) at Rapid River 35 during this first season. Some 58 000 separator barrels

(9 221 m³) of hydrocarbon liquids were produced, which was less than anticipated because of surface equipment problems and lower than calculated well yields. Some 271 bbl (43.1 m³) of water were produced during the first withdrawal cycle. Water analysis indicated this to be fresh water from drilling operations.

TESTING

Flow testing is performed on injection and withdrawal. Withdrawal flow testing is somewhat hampered by liquid fouling of the equipment, but test results are reasonable. Field shut-ins at the end of injection and withdrawal are performed to check inventory and test the integrity of the reservoir. The field has performed quite satisfactorily in terms of deliverability and pressure integrity.

Gas and liquid sampling (discussed in the Gas Inventory section, page 50) involves the continuous monitoring of hydrocarbons injected and withdrawn by component, as well as by phase.

Compressor efficiency and station fuel use is continuously monitored by computer through the development of in-house and other programs designed to track almost every phase of the operation during injection and withdrawal.

WELL MAINTENANCE

Base down-hole well logs were run in all wells to monitor for future gas leakage problems. Pulsed neutron logs were run in all wells that were logged to the surface during drilling, including the reworked existing wells. Gamma-ray neutron logs were run in all wells, including those where pulsed neutron logs were run. One-third of all the wells were to be relogged with gamma-ray neutron logs each year to monitor for gas migration behind pipe.

A corrosion-monitoring program is being developed which will include the periodic running of down-hole logs designed to identify internal and external corrosion problems. Cathodic protection systems are being installed to help mitigate future corrosion.

A safety valve-testing program has been developed to test the integrity of the safety valves during each injection season and again during each withdrawal season.

Chapter 5

DEPLETED HIGH-PERMEABILITY GAS RESERVOIRS

Bistineau Gas Storage Field
United Gas Pipe Line Company
Ralph E. Laenger
Kerry J. Comeaux

INTRODUCTION

A case study in development of depleted high-permeability gas reservoirs is provided in this chapter by the Bistineau, Louisiana, gas storage field. A detailed study of reservoir and geological data was essential to the plan to convert a gas production field into a gas storage field. Fortunately, the company had access to gas production figures and well histories as well as the cooperation of working operators in the area. Table 3 is a summary of the key characteristics of Bistineau storage field.

Located in the northwest part of Louisiana on the east side of Lake Bistineau in western Bienville Parish, the Bistineau gas storage field suits the demands for a flexible set of parameters. Projection of needed gas supplies and deliverability for the Tyler, Shreveport, and Monroe districts from 1961 to 1971 indicated a significant increase. The following two options were considered; the advantages and disadvantages of each are also listed:

- Option 1 – combination of the enlargement of South Louisiana to North Louisiana 30-in. (76.2-cm) facilities and the development of a large underground gas storage field for the Tyler, Shreveport, and Monroe Districts
- Option 2 – the enlargement of South Louisiana to North Louisiana 30-in. (76.2-cm) facilities including extension and enlargement of facilities delivering gas to such 30-in. line

TABLE 3
Data Summary for Bistineau Storage Field

Storage reservoir	Pettet limestone at 5 250 ft (1 600.2 m)
Storage area	12 043 acres (4 873.7 ha)
Proven acres	4 831 acres (1 955.1 ha)
Acre feet	136 970 acres ($1.689\,5\times10^8$ m^3)
Average thickness	28.35 ft (8.6 m)
Average porosity	19.83 percent
Average connate water saturation	15 percent
Initial surface shut-in pressure	2 180 psia (15 031 kPa)
Initial recoverable wet gas	138 160 M^{2}cf ($3.912\,7\times10^9$ m^3)
Cumulative wet gas produced	133 939 M^{2}cf ($3.793\,2\times10^9$ m^3)

OPTION 1

Advantages

- Additional deliverability for periods of heavy demand
- Higher load factor for better performance of gas purchase contract obligations and system operation
- Lower cost of purchasing more gas reserves than actually required to provide adequate deliverability
- Minimized pipeline and compressor additions, both for connection of new reserves and increasing transport capacity to north Louisiana
- Greater operating flexibility
- Advantageous to all districts, with relatively inexpensive increased capacity
- Less costly than Option 2 while providing identical increased deliverabilities

Disadvantages

- Additional gas would be needed by the 1964–65 season. Appreciable time was needed for acquiring storage rights, reworking and drilling wells, acquiring sites and rights-of-way, and constructing facilities to inject cushion gas.

OPTION 2

Advantages

- Additional deliverability for periods of heavy demand
- Greater operating flexibility

Disadvantages

- Continuing purchase of additional reserves required to provide needed additional deliverability
- Continuing increase in pipeline and compressor facilities required for connecting new reserves and providing additional transport capacity to north Louisiana.
- No improvement to system load factor
- Higher investment compared to Option 1

After both options had been weighed, Option 1 was chosen and the conversion of the gas producing field to storage began in late 1965.

REGULATORY REQUIREMENTS

Storage of interstate natural gas requires an application with the FERC (then FPC). The original FPC application was filed on October 29, 1965, and approval came on April 29, 1966. Table 4 provides a summary of all FPC filings.

LAND AND RIGHT-OF-WAY

Acquisition of gas storage rights was studied and reviewed for some time before the actual decision to convert the gas field into a storage reservoir. Contacting the landowners and mineral right owners required time-consuming preliminary planning. Each owner was contacted, and compensation was agreed upon and paid. Also, working interests and wells had to be purchased from various companies.

Generally, 3-acre (1.2-ha) sites were purchased from landowners to drill or work over a well, and the wells were spaced on approximately 40 acres (16.2 ha).

TABLE 4
Summary of FPC Filings for Bistineau Storage Field

	Winter Season Year	Delivery Capacity March 31 MMCF	Number of Wells Observation Old	Number of Wells Injection-Withdrawal Old	New	Total	Compressor Station Horsepower	Dehydration Plant Capacity MMCF	Authorized Maximum Inventory of Gas Stored—MMCF Injected Gas Cushion	Working	Total	Native Gas	Total Gas
FPC DOCKET		$(10^6$ cu					(kw)	$(10^6$ cu	$(10^6$ cu	$(10^6$ cu	$(10^6$ cu	$(10^6$ cu	$(10^6$ cu
Original Application		meters)						meters)	meters)	meters)	meters)	meters)	meters)
Filed: October 29, 1965													
Approved: April 29, 1966		(3.54)					(8 948.4)	(14.86)	(708.00)	(84.96)	(792.96)	(566.40)	(1 359.36)
PHASE I	1966-67	125	9	12	20	32	12 000	525	25 000	3 000	28 000	20 000	48 000
		(8.49)					(8 948.4)	(14.86)	(1 161.12)	(594.72)	(1 755.84)	(566.40)	(2 322.24)
PHASE I	1967-68	300	9	12	20	32	12 000	525	41 000	21 000	62 000	20 000	82 000
		(11.32)					(8 948.4)	(14.86)	(1 387.68)	(764.64)	(2 152.32)	(566.40)	(2 718.72)
PHASE II	1968-69	400	9	12	25	37	12 000	525	49 000	27 000	76 000	20 000	96 000
		(14.16)					(10 439.8)	(14.86)	(1 500.96)	(849.60)	(2 350.56)	(566.40)	(2 916.96)
PHASE III	1969-70	500	9	12	30	42	14 000	525	53 000	30 000	83 000	20 000	103 000
FPC DOCKET													
Filed: July 14, 1967													
Approved: March 11, 1968		(16.9)					(13 422.6)	(14.86)	(1 500.96)	(849.60)	(2 350.56)	(566.40)	(2 916.96)
PHASE III, REVISED 1	1969-70	600	9	12	30	42	18 000	525	53 000	30 000	83 000	20 000	103 000
		(19.82)					(13 422.6)	(14.86)	(1 585.92)	(764.64)	(2 350.56)	(566.40)	(2 916.96)
PHASE IV	1970-71	700	6	15	35	50	18 000	525	56 000	27 000	83 000	20 000	103 000
FPC DOCKET													
Filed: January 27, 1969													
Approved: April 22, 1969		(19.82)					(13 422.6)	(29.73)	(1 585.92)	(1 217.76)	(2 803.68)	(566.40)	(3 370.08)
PHASE III, REVISED 2	1969-70	700	6	15	35	50	18 000	1 050	56 000	43 000	99 000	20 000	119 000

(continued)

TABLE 4 (continued)
Summary of FPC Filings for Bistineau Storage Field

	Winter Season Year	Delivery Capacity March 31 MMCF	Number of Wells				Compressor Station Horsepower	Dehydration Plant Capacity MMCF	Authorized Maximum Inventory of Gas Stored—MMCF				
			Observation Old	Injection-Withdrawal Old	New	Total			Injected Gas Cushion	Working	Total	Native Gas	Total Gas
PHASE IV, REVISED	1970-71	(22.65) 800	6	15	41	56	(13 422.6) 18 000	(29.73) 1 050	(1 585 92) 56 000	(1 217.76) 43 000	(2 803.68) 99 000	(566.40) 20 000	(3 370.08) 119 000
Partner's Portion		(2.83) 100							(198.24) 7 000	(226.56) 8 000	(424.80) 15 000	(70.80) 2 500	(495.60) 17 500
Company's Portion		(19.82) 700							(1 387.68) 49 000	(991.20) 35 000	(2 378.88) 84 000	(495.60) 17 500	(2 874.48) 101 500
FPC DOCKET *System Reinforcement* Filed: December 5, 1969 Approved: May 5, 1970 PHASE V	1971-72	(25.48) 900	6	15	43	58	(13 422.6) 18 000	(29.73) 1 050	(1 727.52) 61 000	(1 217.76) 43 000	(2 945.28) 104 000	(566.50) 20 000	(3 511.68) 124 000
PHASE VI	1972-73	(28.32) 1 000	6	15	45	60	(13 422.6) 18 000	(29.73) 1 050	(1 727.52) 61 000	(1 217.76) 43 000	(2 945.28) 104 000	(566.40) 20 000	(3 511.68) 124 000
Partner's Portion		(2.83) 100							(198.24) 7 000	(226.56) 8 000	(424.80) 15 000	(56.64) 2 000	(481.44) 17 000
Company's Portion		(25.48) 900							(1 529.28) 54 000	(991.20) 35 000	(2 520.48) 89 000	(509.76) 18 000	(3 030.24) 107 000
FPC DOCKET *System Reinforcement* Filed: April 20, 1976 Approved: October 19, 1976		(24.92)					(25 353.8)	(29.73)	(1 478.30)	(1 750.17)	(3 228.48)	(566.40)	(3 794.88)

(continued)

TABLE 4 (continued)
Summary of FPC Filings for Bistineau Storage Field

| | Winter Season | Delivery Capacity March 31 | Number of Wells | | | | Compressor | Dehydration Plant | Authorized Maximum Inventory of Gas Stored—MMCF | | | | |
| | | | Observation | Injection-Withdrawal | | | Station | Capacity | Injected Gas | | | Native | Total |
	Year	MMCF	Old	Old	New	Total	Horsepower	MMCF	Cushion	Working	Total	Gas	Gas
	1976-77	880	5	16	52	68	34 000	1 050	52 200	61 800	114 000	20 000	134 000
		(2.83)							(198.24)	(226.56)	(424.80)	(70.80)	(495.60)
Partner's Portion		100							7 000	8 000	15 000	2 500	17 500
		(22.08)							(1 280.06)	(1 523.61)	(2 803.68)	(495.60)	(3 299.28)
Company's Portion		780							45 200	53 800	99 000	17 500	116 500
STATUS AFTER CERTIFICATE AMENDED													
(June 8, 1979)		(24.92)					(25 353.8)	(46.72)	(1 478.30)	(1 948.41)	(3 426.72)	(566.40)	(3 993.12)
CERTIFICATED:		880	5	16	52	68	34 000	1 650**	52 200	68 800	121 000*	20 000	141 000*
		(2.83)							(198.24)	(226.56)	(424.80)	(70.80)	(495.60)
Partner's Portion		100							7 000	8 000	15 000	2 500	17 500
		(22.08)							(1 280.06)	(1 721.85)	(3 001.92)	(495.60)	(3 497.52)
Company's Portion		780							45 200	60 800	106 000	17 500	123 500
ESTIMATED ORIGINAL GAS IN PLACE													
Gas Produced—1939-66													(3 794.88)
													134 000
Native Gas Left in Place													(566.40)
													20 000
Estimated Total Gas in Reservoir													(4 361.28)
													154 000

*Subject to maximum surface shut-in pressure of 2 175 psia (14 996 kPa).
**Rated at 1 200 psia 100°F continuous operation. (8 274 kPa) @ 38°C

GEOLOGICAL AND ENGINEERING DATA

The field was discovered in March 1921 by the Lakeside Oil Company No. 1 State well, which found shallow gas production in the Tokio Formation [1 950 to 2 000 ft (594 to 610)] and Eagleford Formation [2 500 ft (777 to 793m)] of the Upper Cretaceous Age. Later, deeper drilling discovered gas production in the Pettet Formation of the Lower Cretaceous Age [5 137 to 5 235 ft (1 566 to 1 596 m)], the first completion in which was the DeSoto Oil and Gas Company (R. O. Roy) No. 1 Dunn and Olsen well in January 1937. A total of 19 producing wells and five dry holes were drilled to the Pettet Formation, which was selected as the gas storage reservoir. In addition to the Tokio, Eagleford, and Pettet discoveries, a few later attempts to establish deeper production in the Travis Peak and Cotton Valley formations were not successful.

From 1937 to 1966, the field produced a total of 133.939 Bcf $(3.7932 \times 10^9 \text{ m}^3)$ of gas. At that time, the field was essentially depleted; the several remaining wells were producing small amounts of gas, and the estimated remaining gas reserve was 4.221 Bcf $(1.1954 \times 10^8 \text{ m}^3)$.

Initially, the Bistineau gas storage field was a dry gas reservoir with wells of high deliverability and total recoverable liquids of 5 to 6 bbl/Mcf $(28.1 \text{ to } 33.7 \text{ m}^3/10^9 \text{ m}^3)$ of gas. The production decline curve (see Fig. 26) is almost a straight line and from all engineering aspects an almost textbook example of what a storage field should be. Natural gas is easily injected by the storage plant with five Clark compressors totaling 34 000 hp (25 354 kW) and withdrawn by free flow or compression, with the withdrawal rates as high as 1.2 Bcf $(3.3984 \times 10^7 \text{ m}^3)$ for a 24-hour period.

The Pettet Reservoir of this field is a unique and nearly perfect storage reservoir (see Fig. 27 for a structure map). It is a faulted anticline with the major self-sealing fault running northwest and southeast 400 ft (122 m) down-thrown (east part of field). The northern, western, and southern portions of the structure are bounded by encroachment water and permeability barriers that serve as a seal to prevent any gas loss, and these are down-dip. The encroached water boundaries had moved very slowly over the production life (27 years) of the field and infiltrated only part of the reservoir. The field has the theoretical ability to withdraw in a 5 month period the same quantity of gas produced over a 27 year period. The reservoir mechanism is a pressure-volume relationship; the more gas injected, the higher the pressure.

The top of the structure is covered by the Three Finger Lime (part of the Sligo Formation), which is a dense lime, impermeable and approximately 50 ft (15 m) thick. Each well that has penetrated this impervious mantle has been verified by a cement bond log.

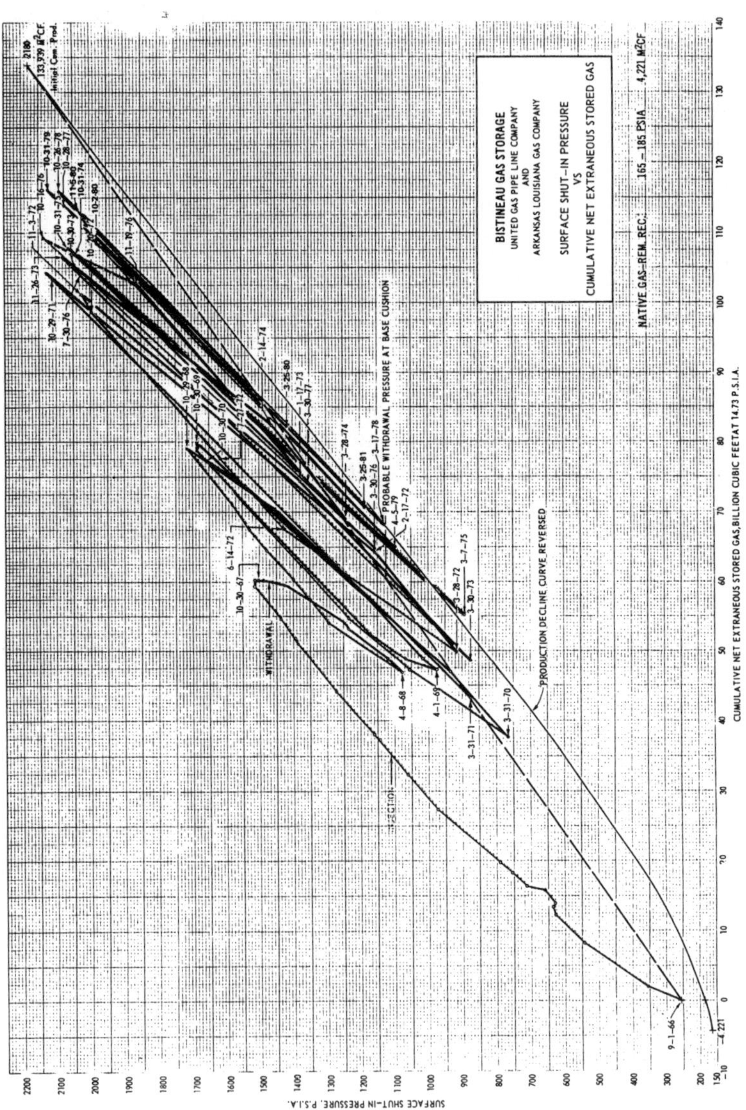

Figure 26. Production decline curve of the Bistineau gas storage field.

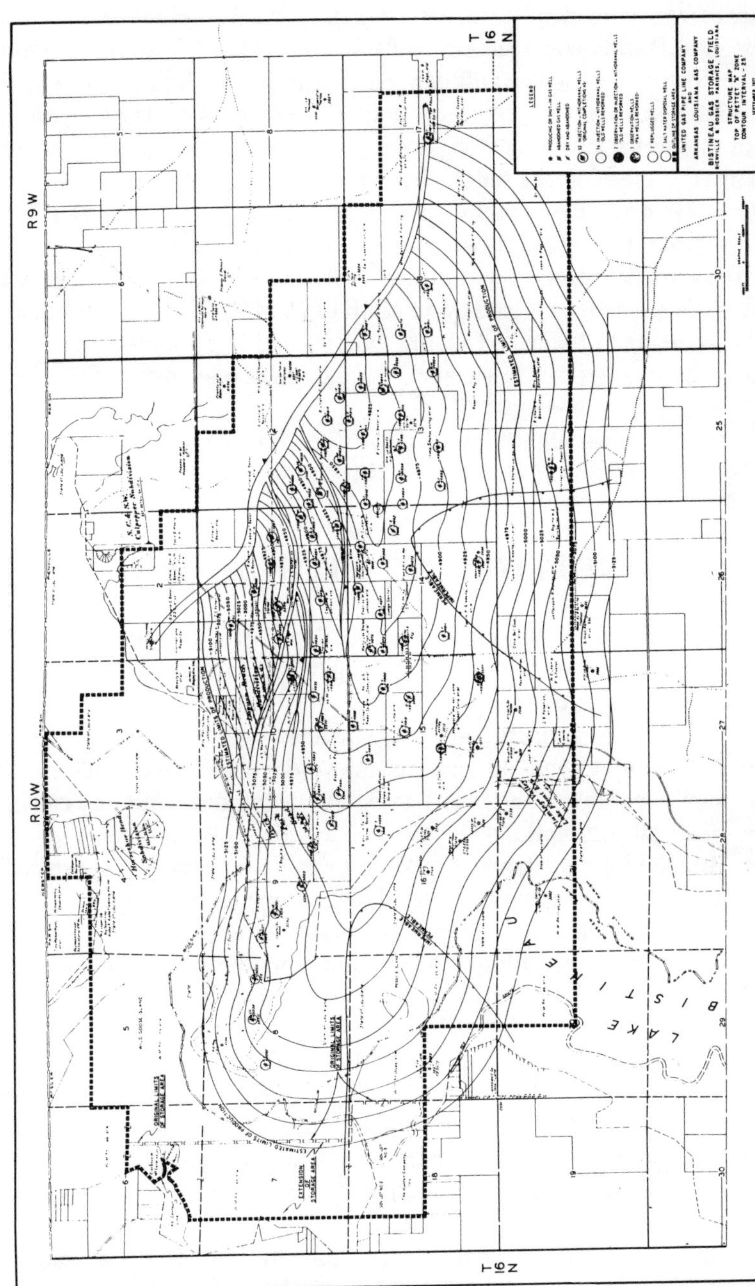

Figure 27. **Structure map of Bistineau gas storage field.**

When the field was abandoned in 1965, the bottom-hole pressure was 180 psia (1 241 kPa). The Pettet Formation is composed of three separate zones (zones A, B, and C) of porosity, which are oolitic limestones. During construction, wells were completed in all three zones. The northwest-southeast fault delineates the eastern limits of the field. The minor east-west adjustment faults do not separate the reservoirs but fault the various zones of porosity in juxtaposition to each other, thus interconnecting the reservoirs across the various fault planes. Approximately 4 831 acres (1 955 ha) are proved, with the storage rights to 12 043 acres (4 874 ha) being the total storage area.

There are 68 injection-withdrawal wells and 5 observation wells, and many wells are characterized by an MER of 50 MMcf/d (1.416×10^6 m³/d). Some water [0.25 bbl/MMcf (1.403 cm/10^6 m³)] is produced whenever volumes in storage reach the certificated cushion at 52.2 Bcf (1.478×10^9 m³).

Designed deliverability capacity with 52.2 Bcf (1.478×10^9 m³) in storage is a total of 880 MMcf/d (2.492×10^7 m³/d).

RECONDITIONING OF OLD WELLS

Following is a step-by-step account of how the plugged and abandoned wells were reconditioned. See Fig. 28 for an illustration of how remedial work was done.

- Tied casing strings back into surface.
- Ran cement bond log, gamma-ray neutron and casing inspection logs to establish and verify integrity of cement bond and casing.
- Used oil-base mud.
- Worked four workover rigs 24 hours a day for fishing operations (one drilling supervisor and one tool pusher per rig).
- Recovered as much tubing as possible from the well.
- Ran cement bond log to verify good bonding in the caprock at approximately 5 000 ft (1 524 m). Ran gamma-ray neutron and casing inspection logs. Squeezed if bonding was not adequate.
- Set a cast-iron bridge plug inside the 5½-in. (14-cm) casing and dumped cement on top of the plug opposite the caprock so that in case the hole was lost (junked), the well would be plugged.
- When it was certain that the well was properly plugged on the bottom, both inside and outside the casing, started work on top of the hole.
- Ran 7-in. (17.8-cm) wash pipe with tungsten carbide shoe, washing over 5½-in. casing until the shoe made no footage.

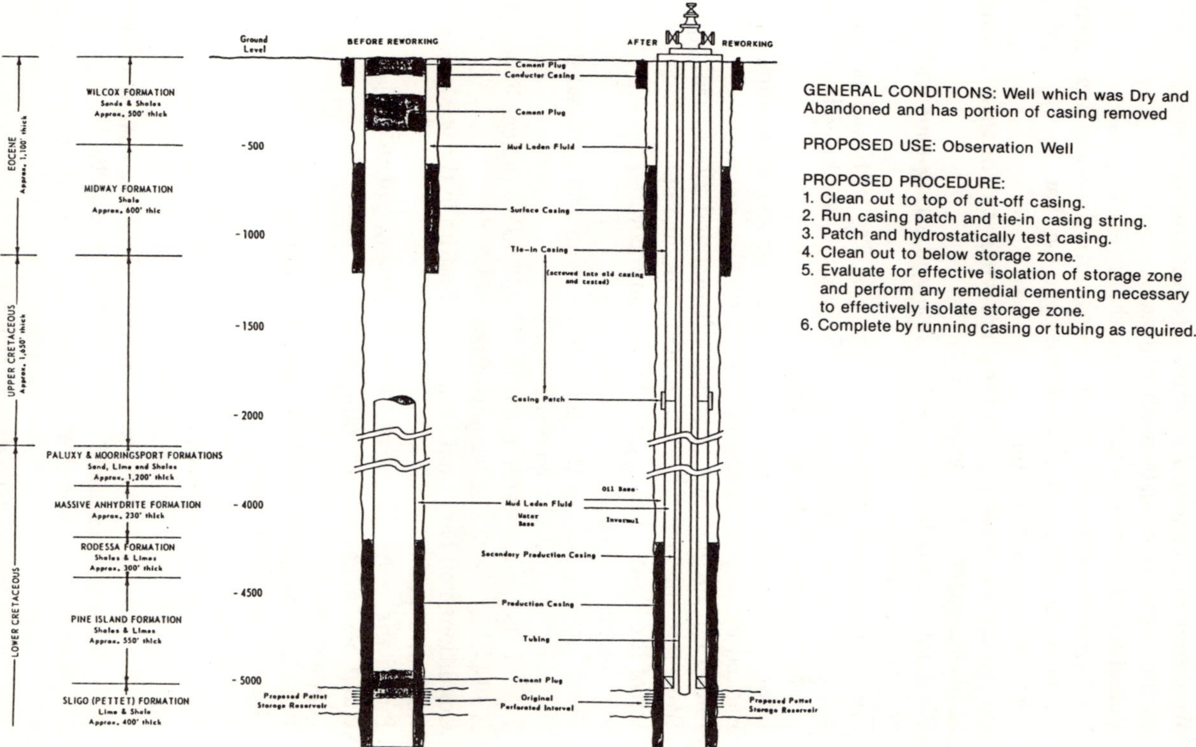

Figure 28. **Remedial work in the Pettet Formation of Bistineau gas storage field.**

- Ran 2⅞-in. (7.3-cm) drill pipe inside 5½-in. casing and set it on solid footing inside the 5½-in casing. Ran 2⅞-in drill pipe to the surface for lineup string.
- Pulled the full string of wash pipe out of the hole and replaced it with tungsten carbide shoe.
- Went back in hole with 7-in. wash pipe and tungsten carbide shoe to the depth previously washed over.
- Pulled 2⅞-in. lineup string out of the hole.
- Continued to wash over 5½-in. casing until the operation reached the point established by cement bond log that there was good bond or good casing point.
- Picked up 7-in. wash pipe to the point where two or three collars of 5½-in. casing were exposed.
- Went into the hole with electric line and string shot and shot selected collar.
- Went into the hole with 2⅞-in. lineup string with back-off tool, and set it inside 5½-in. casing to the point above the collar where the string shot had been run in.
- Lowered 7-in. casing to the previously washed-over point.
- Backed off 5½-in. casing at the string shot point.
- Pulled out of the hole with back-off tool and 2⅞-in. drill pipe.
- Went into the hole with 2⅞-in. drill pipe and spear, caught the top of the 5½-in. casing, pulled fish, and recovered the 5½-in. casing out of the hole.
- Went into the hole with 5½-in. new casing and screwed it into the old casing at the back-off point.
- Tested the casing with 1 500 lb (10 342 kPa); pulled 40 000 lb (18 144 kg) above the weight of the pipe.
- Conditioned oil-base mud around the bottom of the wash pipe at wash-over point.
- Pulled the worn shoe and the wash pipe out of the hole.
- Hung 5½-in. casing in bradenhead.
- Nippled up.
- Went into the hole with 2⅞-in. drill pipe and rock bit and displaced oil-base mud with water.
- Drilled inside the 5½-in. casing, cement, and cast-iron bridge plug.
- Ran wireline packer with position no. 1 nipple below the packer and converter on top of the packer.
- Went into the hole with 3½-in. (8.9-cm) tubing and completed well with overshot receptacle on the bottom of the 3½-in. tubing.

Other observations noted at the time of reconditioning were as follows:

- This is a typical washover and fishing job.

- As much as 3 945 feet (1 202 m) of 7-in. (17.8-cm) wash pipe was run at one point.
- No observation wells per se were drilled.
- The total number of reconditioned wells was 23: 16 as injection-withdrawal wells, 5 as observation wells, and 2 plugged and abandoned. One of the two wells plugged and abandoned was junked; it was plugged properly before fishing operations started.
- Old wells were reworked as observation wells because of their locations.
- The time from first workover to last workover was 9 months.

DRILLING NEW WELLS AND DEFINING FIELD LIMITS

The detailed procedures used in drilling new injection-withdrawal wells were as follows. See Fig. 29 for an illustration of the construction of new wells and Fig. 30 for a map of the storage field.

- Wells were generally on 40-acre (16.2-ha) spacing, with surface locations taken into consideration.
- The main concept in drilling was to use large casing to permit high deliverability rates.
- The first gas injection occurred on September 1, 1966.
- Twenty new injection-withdrawal wells were drilled.
- Water-base mud was used.
- A 24-in. (61-cm) hole was drilled to 400 ft (122 m).
- A string of 20-in. (50.8-cm) casing was set at 400 ft and cemented to the surface with common cement plus 16 percent gel and 5 percent salt. The tailing cement was common cement plus 2 percent calcium chloride.
- A 17½-in. (44.5-cm) hole was drilled to 1 400 ft (427 m).
- A string of 13⅜-in. (34-cm) casing was set at 1 400 ft and cemented to the surface with common cement plus 16 percent gel and 5 percent salt, 50–50 Pozmix A cement, and common cement plus 2 percent calcium chloride.
- A 12¼-in. (31.1-cm) hole was drilled to 5 000 ft (1 524 m) approximately, or to the middle of the third finger of the Three Finger Lime of the Sligo Formation. Electric log surveys were run to 5 000± ft.
- A string of 9⅝-in. (24.4-cm) casing was set at 5 000± ft and cemented up into the surface casing at 1 400 ft, using a stage tool at approximately 3 600 ft (1 097 m).

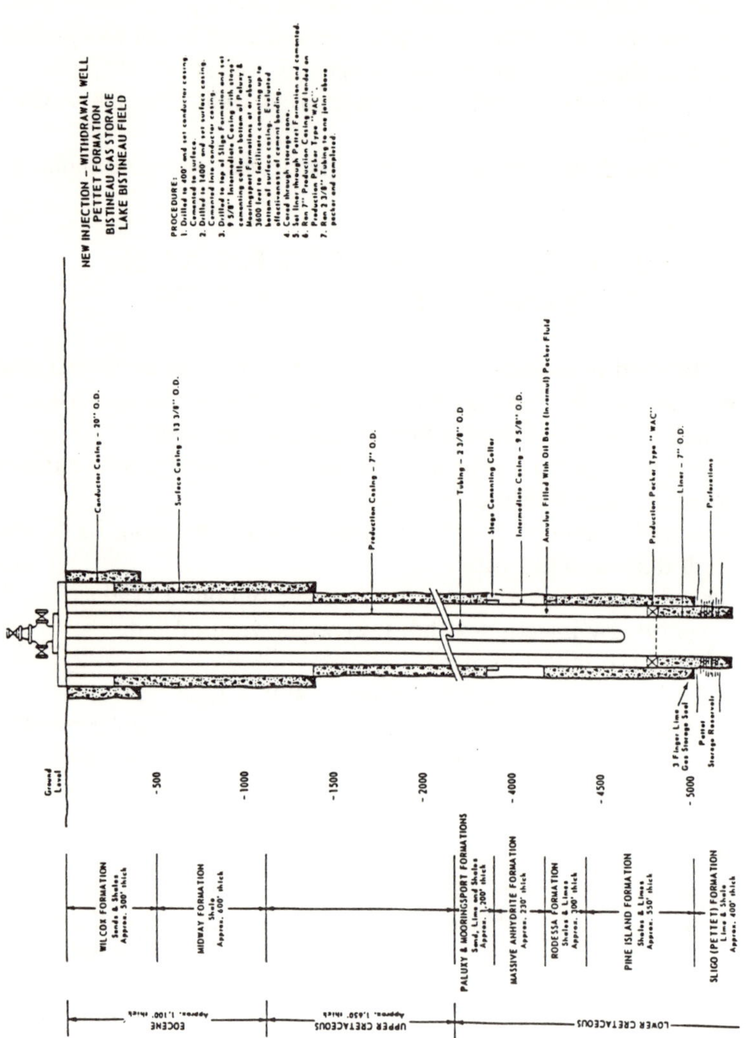

Figure 29. Construction of new wells in Bistineau gas storage field.

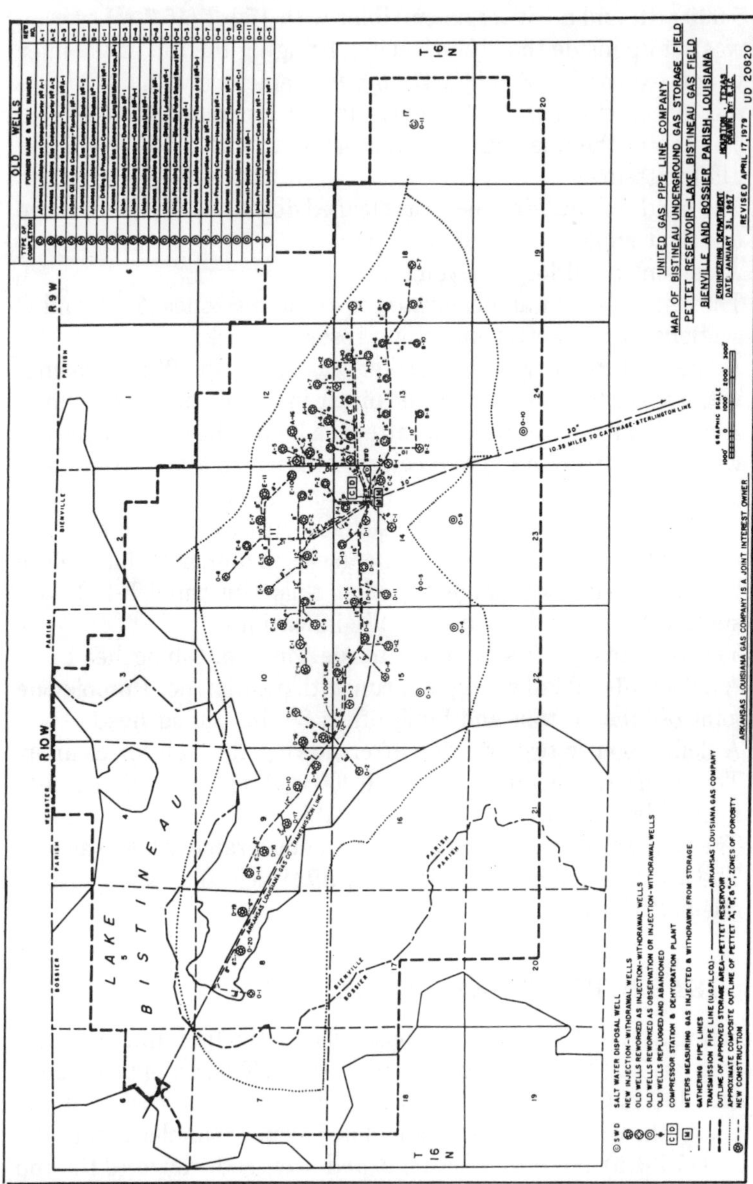

Figure 30. Map of Bistineau gas storage field.

- The bonding in the caprock (Three Finger Lime) was verified with a cement bond log, and a gamma-ray neutron log was run to the surface.
- Mud was displaced with water. A float shoe and collar shoe were drilled out in the 9⅝-in. casing and an 8¾-in. (22.2-cm) hole was drilled to 5 400 ft (1 646 m). Electric log surveys were run to 5 000± ft, and a 7-in. (17.8-cm) liner with 150-ft (45.7-m) overlap was set up inside the 9⅝-in. casing, bringing the top of the liner to approximately 4 850 ft (1 478 m), and cemented with 50–50 Poz-mix A cement plus 12 percent salt and 1 percent CFR-2.
- The top of the liner was tested at 4 850 ft with 2 500 lb (17 230 kPa) pressure.
- The inside of the 7-in. liner was cleaned down to 5 ft (1.5 m) above the float shoe.
- A cement bond log was run.
- The 7-in. casing was perforated in reservoir zones A, B, and C sections of the Pettet limestone reservoir rock.
- A special type of wireline packer was set in the 9⅝-in. casing, with packer tail pipe and profile nipple in the top of the 7-in. liner.
- A position no. 1 X nipple was attached below the wireline packer with a converter on top to receive a 7-in. overshot-type receptacle (flotation seals to allow for expansion and contraction).
- The 7-in. casing was run to the surface with a 7-in. overshot-type receptacle over converter of packer (seals in overshot). Just before the 7-in. string was landed, packer fluid (Invermul F-100) was spotted between the 7-in. and 9⅝-in. casings.
- The 7-in. casing was hung off inside the 7-in. tubing head.
- A 2⅜-in. (6-cm) kill string was run with position no. 1 nipple one joint off the bottom and hung off in 2⅜-in. tubing head.
- A deliverability test was run after flowing the well for cleanup.
- The well was acidized with 20 000 gal (76 m³) 28 percent hydrochloric acid.
- The well was flowed for cleanup and a deliverability test was run. The average deliverability increase after acidizing was two-and-one-half times.

WELLHEAD STRUCTURES AND ANNULI

Starting at the 13⅜ in. (34-cm) bradenhead, the first annulus is between the 9⅝ in. (24.4-cm) and 13⅜ in. casings. This annulus extends down to 1 400 ft (427 cm) to the bottom of the 13⅜-in. casing or top of cement in the 9⅝-in. by 13⅜-in. annulus. The next annulus is between the 7-in. (17.8-cm) and 9⅝-in. casings, and it extends down to the top of the packer at approximately 5 000 ft (1 524 m). The special packer is set in the 9⅝-in. casing near the top of the 7-in. liner with the 5½-in. (14-cm) outside diameter (OD) tail pipe X nipple and the nipple spacer set down in the 7-in. liner. This method provides a way of setting a plug in the position no. 1 nipple below the packer, working on just about any mechanical tool failure above the packer, then pulling the plug and bringing the well in without risk of disturbing or decreasing the deliverability of the well because of fluid loss. The 7-in. casing and the 2⅜-in. (6-cm) tubing are both gas strings. The gas mostly flows through the 7" casing (17.8 cm). The 2⅜" (6 cm) tubing is used for a kill string and to lubricate chemical corrosion inhibitor into the well. This type of wellhead structure (see Fig. 29) permits maximum flow of a well without any encumbrances of unnecessary equipment.

At the surface there are two 4-in. (10.2-cm), 3 000 lb (20 680-kPa), working pressure valves, one on each side of the 7-in. (17.8-cm) annulus, connected to expansion loops, and then to a tee going into the 6-in. (15.24-cm) meter run. Two 2-in. (5.1-cm), 3 000 lb working pressure valves control the 2⅜-in. (6-cm) tubing.

There are no separation facilities at the well installation; all separation takes place at the plant.

RESERVOIR PERFORMANCE AND DELIVERY CAPACITY

All of the design and development at the gas storage field has been based on the relation of required delivery capacity (fixed), required working gas (fixed), compressor and dehydration plant (fixed), number of wells (variable), cushion gas (variable). It was always recognized that the number of wells and the volume of cushion gas were variables subject to adjustment to meet the other fixed requirements of design and time. Additional wells were drilled and the volume of the cushion gas was adjusted to meet the changing demands for deliverability and working gas for the system. Those demands are summarized in Table 5.

The requirements for the system for the 1968–69, 1969–70, and 1970–71 winters resulted in a maximum volume in storage of about 80 Bcf (2.266×10^9 m³) at the end of the injection cycle; volumes withdrawn were 33, 43, and 37 Bcf (0.9346×10^9 m³, 1.218×10^9 m³ and 1.048×10^9 m³), respectively. These similar operations resulted in a partial equilibrium and a reasonably stabilized reservoir performance for these conditions. Very high delivery capacities were obtained with reasonably low volumes of cushion gas because the total capacity of all of the reservoir was not being used and the storage gas tended to remain near the

TABLE 5
Summary of Gas Storage Utilization for Bistineau Gas Storage Field

Year	Maximum BCF in Storage (10^9 m^3)	Minimum in Storage (10^9 m^3)	MMCF Cushion (10^6 m^3)	Designed Deliverability (10^6 m^3)	No. of Wells
1971–72	104.29 (2.953 5)	48.81 (1.382 3)	61 000 (1 727.5)	900 (25.49)	58
1972–73	101.34 (2.869 9)	48.49 (1.373 2)	61 000 (1 727.5)	1,000 (28.32)	60
1973–74	106.52 (3.016 6)	69.24 (1.960 9)	61 000 (1 727.5)	1,000 (28.32)	60
1974–75	106.84 (3.025 7)	53.09 (1.503 5)	61 000 (1 727.5)	1 000 (28.32)	60
1975–76	107.92 (3.056 3)	64.45 (1.825 2)	61 000 (1 727.5)	880 (24.92)	60
1976–77	109.34 (3.096 5)	66.06 (1.870 8)	52,200 (1 478.3)	880 (24.92)	68
1977–78	114.02 (3.229 0)	66.12 (1.872 5)	52 200 (1 478.3)	880 (24.92)	68
1978–79	117.07 (3.315 4)	63.47 (1.797 5)	52 200 (1 478.3)	880 (24.92)	68
1979–80	117.12 (3.316 8)	78.96 (2.236 1)	52 200 (1 478.3)	880 (24.92)	68

top of the structure, where most of the wells are located. These reasonably stabilized operating conditions were used for estimating long-range future operating parameters such as number of injection-withdrawal wells and requirements for cushion gas and working gas. The estimates made in 1969 and approved by the FPC on May 5, 1970, were used as the basis for construction and expansion in 1973.

Using the total reservoir at maximum capacity will result in more gas in the thinner, less permeable areas on the periphery of the reservoir where few operating injection-withdrawal wells exist. During the 1973–74 winter the United Gas Pipeline portion of the storage field was operated at high delivery capacity for peaking. Gas was injected and withdrawn in November and December 1973, and consequently the storage companies had a total volume in storage of 101.47 Bcf ($2.873\,6 \times 10^9$ m^3) on January 1, 1974. This volume maintained a high pressure on the top of the structure for a long time and resulted in pushing about 3 or 4 Bcf (8.496×10^7 m³ or 1.1328×10^8 m³) of gas farther out on the periphery of the structure, where it is not readily accessible to the injection-withdrawal wells. During the next 3 months of that withdrawal cycle the reservoir was pulled down to only about 70 BCF (1.9824×10^9 m³). For several months gas was pushed farther out on the structure and did not create any differential or time for much, if any, of it to move back to the top portion of the structure. This type of operation allows the injection of more gas into the storage reservoir for a larger volume of working gas by allowing the pressure on the top of the structure to increase slowly, but the lower pressure results in decreased deliverability. The operator must decide whether to place the major emphasis on high deliverability with slightly lower working gas, or slightly lower deliverability with a larger volume of working gas.

STIMULATION

In completing a new well, the procedure taken in construction would be as follows, after setting the packer and installing the Christmas tree:
- Flow the well to surface and clean up.
- Determine flow capacity.
- Acidize with 1 000 gal (3 785 L) 15-percent hydrochloric acid.
- Flow the well to the surface for cleanup.
- Determine flow capacity.
- Acidize with 20 000 gal (75 000 L) 28-percent hydrochloric acid.
- Flow the well to the surface and determine flow capacity.

This type of completion procedure was followed consistently. After the large acid treatment, well deliverability usually increased two-and-one-half times.

GATHERING SYSTEM

The gathering system is composed of six lateral lines (A to F), varying in sizes from 6 in. (15.2 cm), 8 in. (20.3 cm), 10 in. (25.4 cm), 12 in. (30.48 cm) and 16 in. (40.64 cm) (see Fig. 30) with field pressure coming all the way to the plant. Three lateral lines (A, D, and E) are looped; 10 valves are spotted strategically in the lateral lines in case of a line failure. Three field regulators at the plant [two 12-in. (30.5-cm) and one 6-in. (15.2-cm)] decrease maximum pressure of 2 175 psia (14 996 kPa) (field pressure) down to 1 050 psia (7 240 kPa) (plant pressure). This system can handle an MER of 1.2 Bcf/d (3.3984×10^7 m³/d) with 63 Bcf (1.7842×10^9 m³) in storage (net extraneous gas).

Table 6 shows the design load carried by each lateral on peak day design deliverability.

This is a rather large gathering system, and, of course, from reading Table 6 one can see that lateral A has the best wells. These wells are drilled in the east-central part of the field.

COMPRESSION

Compression was 12 000 hp (8 948 kW) from December 1966 to September 1969, when it was expanded to 18 000 hp (13 422 kW). Compression was further expanded to 26 000 hp (19 388 kW) in April 1978 and 34 000 hp (25 354 kW) in May 1978. (See Fig. 31 for an isometric drawing of major gas piping at the storage station.) Bistineau has in storage a certificated cushion of 52.2 Bcf (1.478×10^9 m³), 45.2 Bcf (1.28×10^9 m³) owned by the company, and 7 Bcf (1.982×10^8 m³) owned by the partner. Using compression, this storage field is capable of delivering to the pipelines 880 MMcf/d (2.492×10^7 m³/d) at cushion when theoretical wellhead shut-in pressures are 731 psia (5 040 kPa) and the suction pressure to the compressor is 493 psia (3 300 kPa).

On injection, with 74 Bcf (2.096×10^9 m³) in storage, 1 330 psia (9 170 kPa) shut-in pressure, 600 psia (4 137-kPa) suction pressure, five engines, and 34 000 hp (25 354 kW), 665 MMcf/d (1.883×10^7 m³/d) can be injected. With 700-psia (4 826-kPa) suction pressure, 803 MMcf/d (2.274×10^7 m³) can be injected.

TABLE 6
Contribution of Laterals to Field Gathering System

Lateral Line	No. of Wells	Percentage of Contribution	Approx. Volume Rate MMcf/d (10^6 m³)
A	16	38	456 (12.914)
B	11	13	156 (4.418)
C	2	2	24 (.680)
D	22	16	192 (5.437)
E	15	27	324 (9.176)
F	2	4	48 (1.359)
Total	68	100%	1 200 MMcf/d MER (33.984× 10^6m³/d)

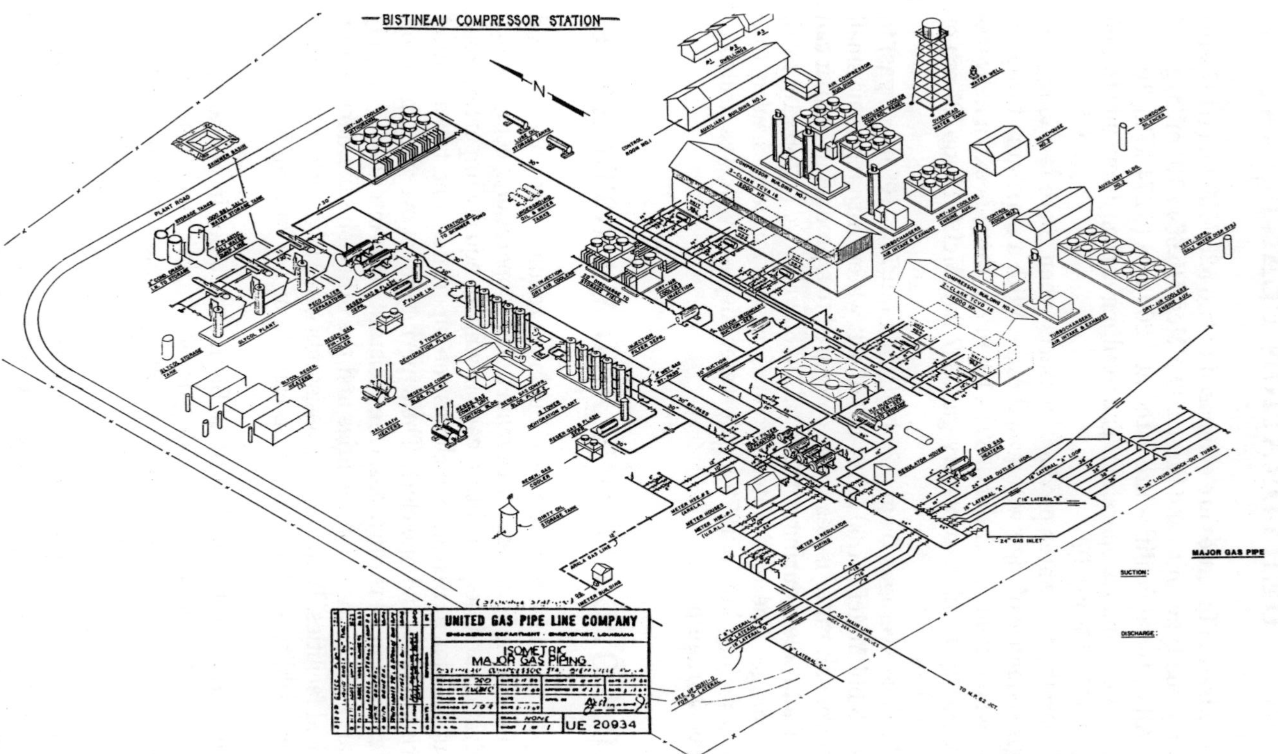

Figure 31. Isometric drawing of major gas piping at Bistineau storage.

DEHYDRATION PLANT

Dehydration facilities are composed of two 525-MMcf/d (1.487×10^7 m³/d) dry bed type units and three 200-MMcf/d (5.664×10^6 m³/d) glycol units. All of these facilities are rated at 1 200 psia (8 274 kPa), 100 °F (38 °C) for continuous operation. The resulting capacity amounts to 1.65 Bcf/d (4.673×10^7 m³/d).

Generally, salt water production starts when there is a long, continuous sequence of high withdrawal rates. With a maximum flow rate on the whole field of 1.2 Bcf (3.398×10^7 m³/d), there is extra dehydration capacity, which assures dry gas of 7 lb/MMcf (113 kg/10^6 m³) of water, or less.

The field averages from 0.25 bbl/MMcf (39.74 L/10^6 m³ of gas) to about 0.3 bbl/MMcf (47.69 L/10^6 m³) of salt water. Although this small amount of salt water is negligible compared to a producing field, it can become a problem during withdrawals of 800 MMcf/d (2.266×10^7 m³/d) from storage. The salt water production was formerly hauled to a disposal facility, but in 1979 a salt water disposal well was drilled at the gas storage plant.

DAILY OPERATIONS AND MAINTENANCE

Operating a facility the size of Bistineau requires a considerable and constant effort. There are three main areas of operations: compression, dehydration and field facilities, each having an equally important role.

Supervision at Bistineau includes one superintendent, two assistant superintendents and one clerk. The superintendent oversees and coordinates work in all three areas. One assistant superintendent is mainly in charge of compression facilities and the other one the dehydration and field facilities. While the two assistants generally work in their area of responsiblity, each one must be capable of performing in all three areas.

The largest volume of work at Bistineau comes in operating and maintaining the compressor and dehydration facilities mentioned earlier. Personnel includes two electronics technicians, one assistant electronics technician, two compressor mechanics, four operators, four compressor helpers, two pipeliners and one mechanical analyst. Bistineau is manned 24 hours a day by one operator who runs compression and dehydration facilities. All other personnel are involved in daily repairs, routine and scheduled maintenance, testing, calibration and inspection of facilities.

The field facilities, which include injection/withdrawal and observation wells, gathering lines, measuring and regulating equipment and saltwater disposal equipment are operated and maintained by two storage technicians and one pipeliner.

No discussion of underground storage would be complete without mentioning inventory verification. An underground storage reservoir the size of Bistineau requires substantial investment of dollars in the form of gas. Accurately keeping up with how much gas is in the ground and making sure it stays there is essential. As mentioned previously, Bistineau is an almost perfect example of a volumetric relationship. Therefore, to keep up with the volume of gas in the reservoir, a plot of pressure versus volume is maintained. Semiannually, the field is shut-in (ranging from 4-24 hours) and each well's pressure is dead-weighted. These pressures are averaged and plotted versus booked inventory (see figure 26). This curve or hysteresis shold track along the reversed decline curve.

However, as the role of United Gas went from merchant to transporter, the role of storage changed from definitive injection and withdrawal cycles to little or no seasonal cycling. Rather, Bistineau is used as a large regulating facility correcting imbalances that occur on the pipeline system. This, on injection one day on withdrawal the next, has caused refinements to be made in the inventory verification process. Now, along with the semiannual shut-ins, monthly pressure checks are made as only certain wells in the field are shut in for a 24-hour period. This helps to balance out the ill effects of the on-again off-again operation.

The following operation procedures are carried out year-round:

- A 24-hour log of each day's events is kept, including such details as volumetric rates, pressures, temperatures, wells used, and work done. This record may be referred to later if necessary, and it may also be used by storage engineers to monitor reservoir performance.
- All flow and pressure charts at the plant are reviewed and changed daily.
- Casing pressure is reviewed weekly on all wells in the field to determine the possibility of packer failure or tubing or casing leaks.
- Maintenance is done on all testing equipment such as manometers, dead weight testers, and thermometers used in testing for and maintaining measurement accuracy.
- The water-filtering system and salt water disposal plant and well are maintained to ensure environmental compliance with all state and federal regulations.

Chapter 6

DEPLETED LOW-PERMEABILITY GAS RESERVOIR

Wharton Field
National Fuel Gas Corporation
Earl L. Welton

INTRODUCTION

For a case study of a low-permeability gas field used as a gas storage reservoir, this chapter describes the Wharton field, which is located in Potter and Cameron countries, Pennsylvania.

GEOLOGY

The Wharton field lies within the Allegheny Plateau Province, where alternating anticlines and synclines trend generally in a northeast-southwest direction. The gas field is located on an elongated dome along the axis of the Marshlands anticline, which is one of the main structural features in the area. When accumulation of gas is found, it is usually in the anticlinal domes, which are often associated with faulting.

The Oriskany sandstone is the principal gas-producing formation in the area. It is a gray to white-colored, fine to coarse-grained, quartzose sandstone. The average sand thickness is approximately 31 ft (9.4 m). The pay zone in the sand averages about 7 ft (2.1 m) and is generally found in the top 12 ft (3.7 m) of the formation.

The field has approximately 250 ft (76.2 m) of structural closure on the Oriskany sandstone. A saddle exists between the main pool and the considerably less permeable East Fork area at the northeast end of the structure. At the southwest or First Fork end of the field, pressure differentials, dry holes, and abnormal formation intervals indicate that faulting separates wells located in the First Fork area from the main portion of the field. Dry holes, salt water, and abnormal formation intervals indicate the presence of a major fault along the southeast flank of the entire structure.

Although Oriskany gas was discovered as early as 1933 in the northeast (East Fork) portion of the field, little gas was removed or the true size of field indicated until March 1948, when the first well was drilled in the main portion of the field. A geological structure map is shown in Fig. 32. This discovery well of the main portion of the field (Well Wh-4 on the map) had an initial open flow of 7 400 Mcf (210×10^3 m³) daily. Active drilling followed, resulting in considerable development especially during the years 1948 and 1949. Rock pressures as high as 3 600 psig (24 822 kPa) were encountered in the main pool of the field. Open flows in the portions of the field averaged approximately 4 000 Mcf/d (113×10^3 m³). One well had an initial open flow of 22 000 Mcf/d (623×10 m³).

Drilling in the entire gas field resulted in 44 productive gas wells and 8 dry holes. The average depth of the wells is 6 049 ft (1 843.7 m), ranging from 5 379 ft to 6 525 ft (1 639.5 to 1 988.8 m). The limits of the field are indicated by the presence of the relatively impermeable, tight sand conditions and dry holes at both the northeast and southwest ends of the fields, by several dry holes on the northwest flank, and by dry holes and the major fault on the southeast flank. Salt water was found in the dry holes on the northwest and southeast flanks of the structure. The field is approximately 10 mi (16.1 km) long and 2 mi (3.2 km) wide at its widest point and contains approximately 8 700 acres (3,523.5 ha).

LAND AND RIGHT-OF-WAY

Production and storage rights were acquired over a period of many years before the area was developed for storage operations. Generally, a lease form was used, which included both production rights and storage rights. The leases were held by paying for the production of native gas until the native gas was depleted to an uneconomical production level, at which time the leases were maintained under storage rights. The storage rights are paid on an annual dollar-per-acre (dollar-per-hectare) basis. If a storage well is located on a lease, the company pays an additional annual fee for the right to maintain and operate the well.

Surface rights for the construction of the main trunk line system and well lines within the pool boundary were part of the leasing agreement. Any easements acquired outside the pool boundary for pipelines were either purchased in fee or paid for on a cost-per-foot (meter) basis plus any damages. The compressor station sites and operating centers were purchased in fee by the companies.

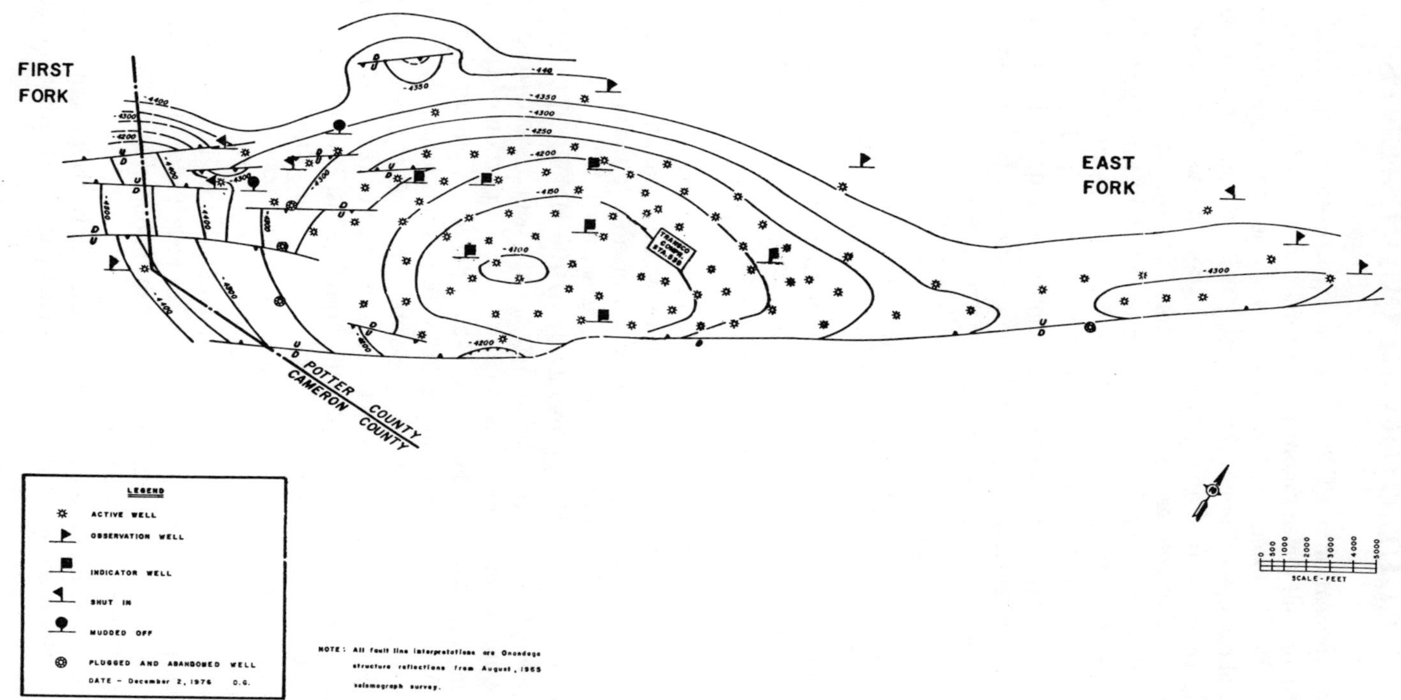

Figure 32. A geological structure map for Wharton storage field.

REGULATORY REQUIREMENTS

The application to FERC for authority to construct was filed early in 1961 for full development of the total capacity [35.0 Bcf (99.1×10^9 m³)] of the reservoir. This application was rejected by FERC because the total top-gas volume was not committed or required by the parties. In July 1962, the first amendment to the docket was filed for a partial development of the reservoir. This application was approved and construction began in early 1963. There were two intermediate development programs; the ultimate program filed for in December 1964 eventually led to the full development of the storage field.

All necessary state and local filings for permits such as well permits, stream crossings, highway crossings, and railroad crossings were filed for and received before any construction started.

GAS INVENTORY

The decision to develop the field for storage of gas was based on many reservoir engineering and geological studies to determine the ability of the reservoir to perform and meet the needs of the companies and its customers.

All gas production figures were obtained either from the company's existing records or from other producers in the area. These data, along with the geological and reservoir data available to the company, were used to verify the actual size of the reservoir.

In November 1958 approximately 31 700 MMcf (897.6×10^6 m³) — base pressure 15.325 psia (105.665 kPa) — had been measured out of the field, approximately 82 percent from the main portion of the field, 8 percent from the East Fork area, and 8 percent from the First Fork area. An estimated 216 MMcf (6.116×10^6 m³) was lost from drilling in and completing the wells. The rock pressure in November 1958 from 16 wells still active indicated an average pressure of 182 psig (1 255 kPa). These data indicated that 31 900 MMcf (903.3×10^6 m³) at 15.325 pressure base were removed from the reservoir while the pressure dropped from 3 600 psig (24 822 kPa) to 182 psig. Converted to a 14.7-psia (101.36 kPa) pressure base, a volume of 31 900 MMcf becomes 33.256 MMcf (941.7×10^6 m³). The Mcf/lb (m³/kPa) decline is 9 730 Mcf/lb (39.96×10^3 m³/kPa). The ultimate total reserves of the field were estimated to be 9 730 Mcf/lb times 3 600 lb (24 822 kPa), or 35 000 MMcf (991.1×10^6 m³) at 14.7 psia base.

Since the first injection of gas, measures have been taken to monitor the storage inventory and account for all gas losses either measured, estimated, or determined unrecoverable from the reservoir or the surrounding area. Unrecoverable losses are determined by monitoring Mcf/lb (m³/kPa) during the injection and withdrawal cycles as well as for the total inventory. No losses to the reservoir or surrounding area have been observed using this technique. All known and estimated losses are written off during the month in which they occur.

DELIVERABILITY

In developing the area as a gas storage pool, a maximum average stabilized wellhead pressure of 3 600 psig (24 822 kPa) was assumed. On the basis of gas production from the area, the pool capacity at 3 600 psig was 35 000 MMcf (991.1 x 10⁶ m³). Of this total capacity, 24 000 MMcf (679.6×10⁶ m³) was to be used as active gas and 11 000 MMcf (311.5×10⁶ m³) as cushion gas. The critical withdrawal period is at a time when 85 percent of all the active gas has been removed. Under this condition the stabilized wellhead pressure is 1 501 psig (10 349 kPa) and a deliverability of 300 MMcf/d (8.5×10⁶ m³/d) was required. The critical injection requirement is 118 MMcf/d (3.34×10⁶ m³/d), and the Wharton design of horsepower and number of wells was based on the last day of injection, when the wellhead stabilized pressure is 3 600 psig. Results of these calculations indicated that the most economical combination to meet these conditions was 88 wells and 8 300 hp (6 189.3 kW).

Original deliverability data on 31 wells in the storage area were obtained and an average well deliverability curve was established at $\Delta\,p^2$ of 10⁶ power for three areas within the pool limits: the main pool area, the First Fork area, and the East Fork area. The average well deliverability curve for the main pool was 1 156 Mcf/d (33×10³ m³/d), for the First Fork area 208 Mcf/d (6×10³ m³/d), and for the East Fork area 11 Mcf/d (0.3×10³ m³/d). (See Figs. 33, 34, and 35.)

After the field was developed and used for storage, it was discovered through back-pressure testing data that there were in fact four different areas of deliverability in the main pool (see Fig. 36).

The results of the back-pressure testing data are shown on the curves, Figs. 37 through 40. Fig. 41 is a composite average well curve for the total field based on actual data collected during operations.

The average well curves based on original well data indicated a deliverability at $\Delta p^2 \times 10^6$ of 1 156 Mcf (33×10³ m³). The average well curve based on actual data collected during operations shows an average well deliverability at $\Delta p^2 \times 10^6$ of 750 Mcf (21×10³ m³). Because of the limited original deliverability data and the low-matrix permeability encountered throughout the main pool area, the performance of this area was slightly overestimated.

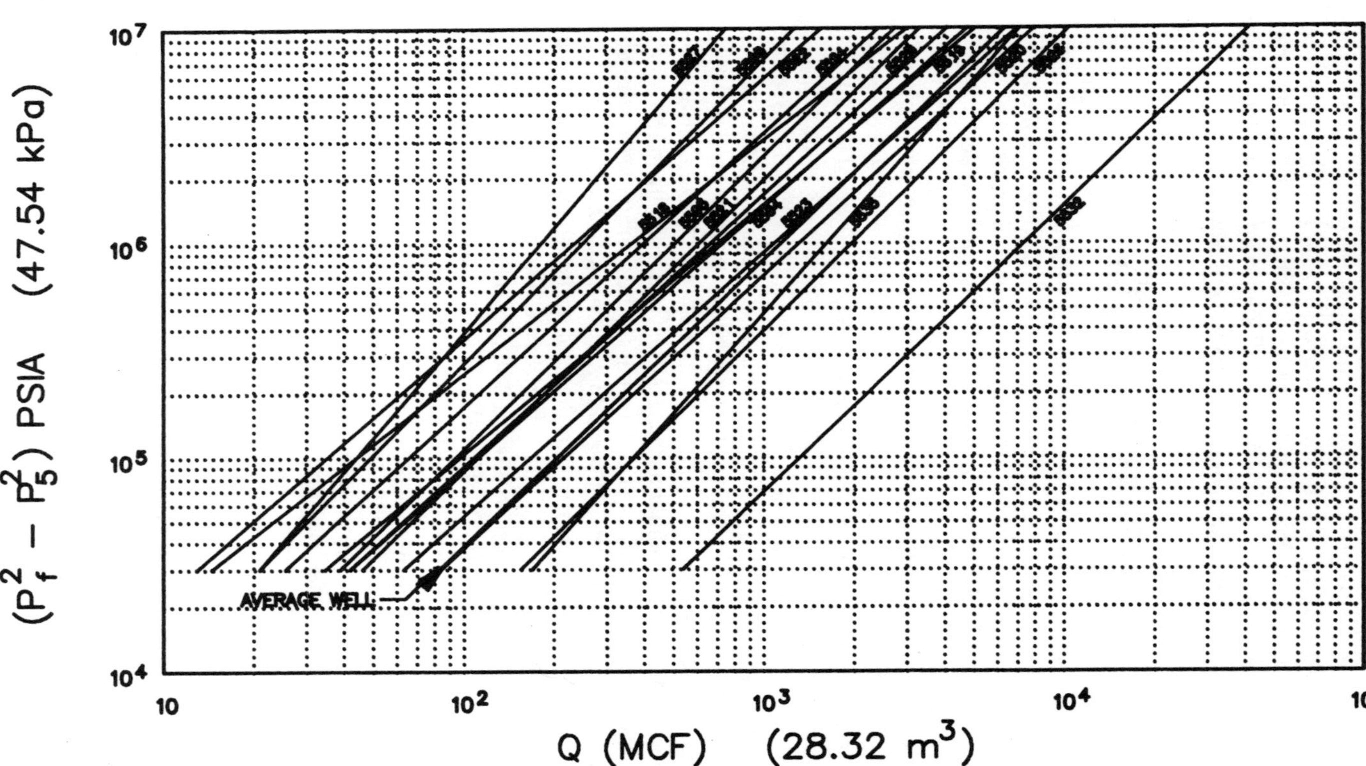

Figure 33. Wharton deliverability curves—main pool area.

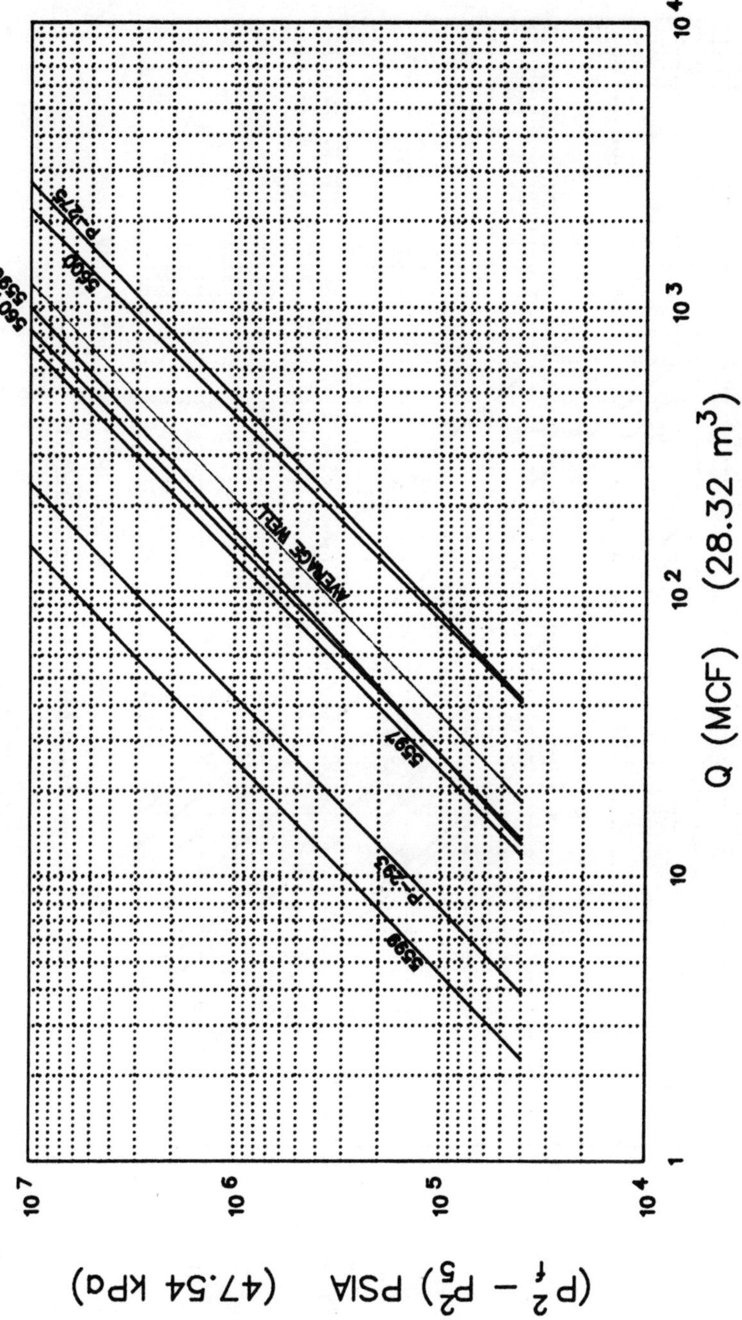

Figure 34. Wharton deliverability curves—First Fork area.

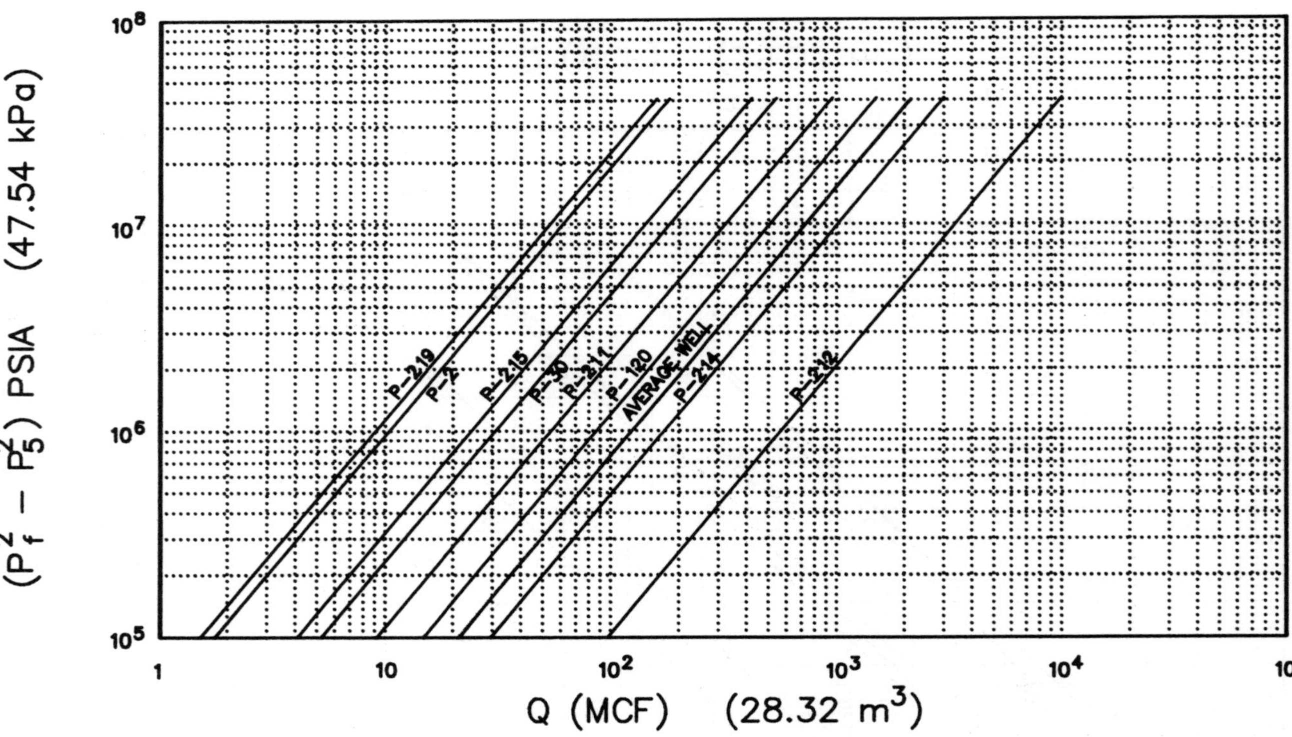

Figure 35. **Wharton deliverability curves—East Fork area.**

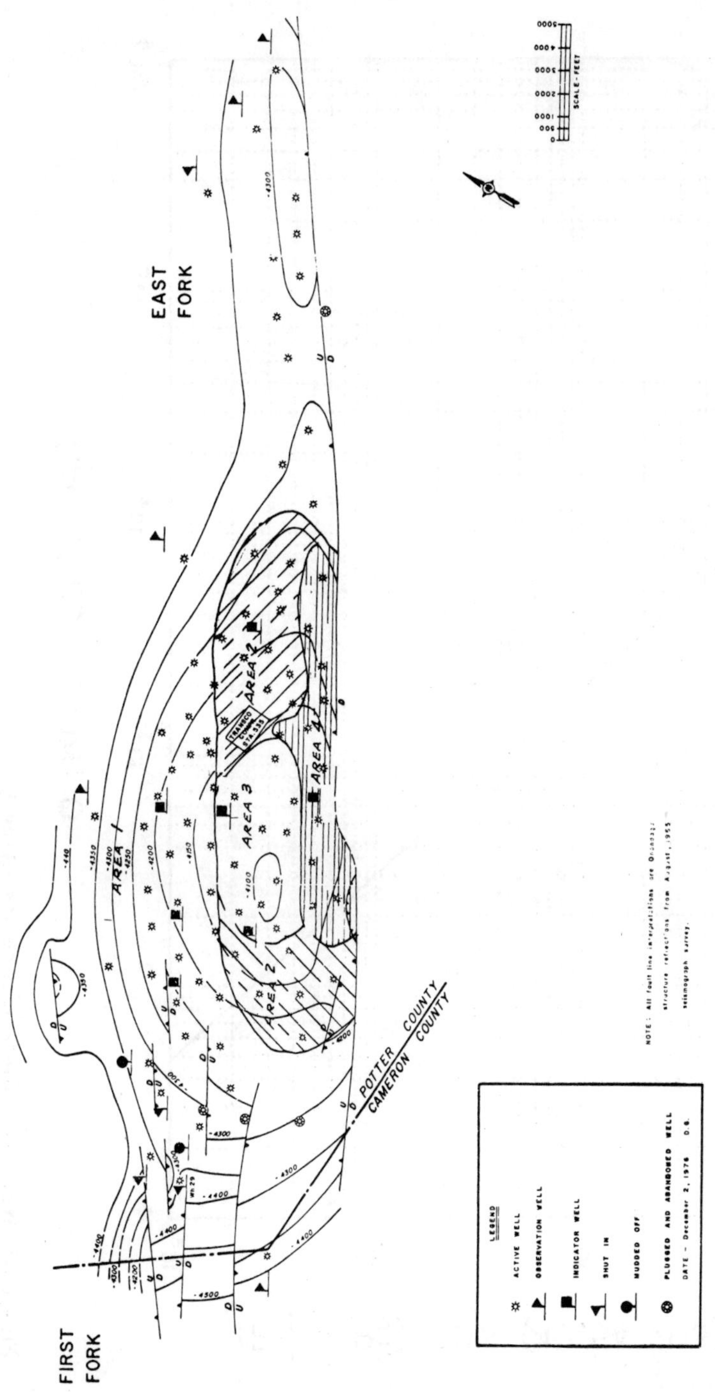

Figure 36. **Wharton deliverability areas.**

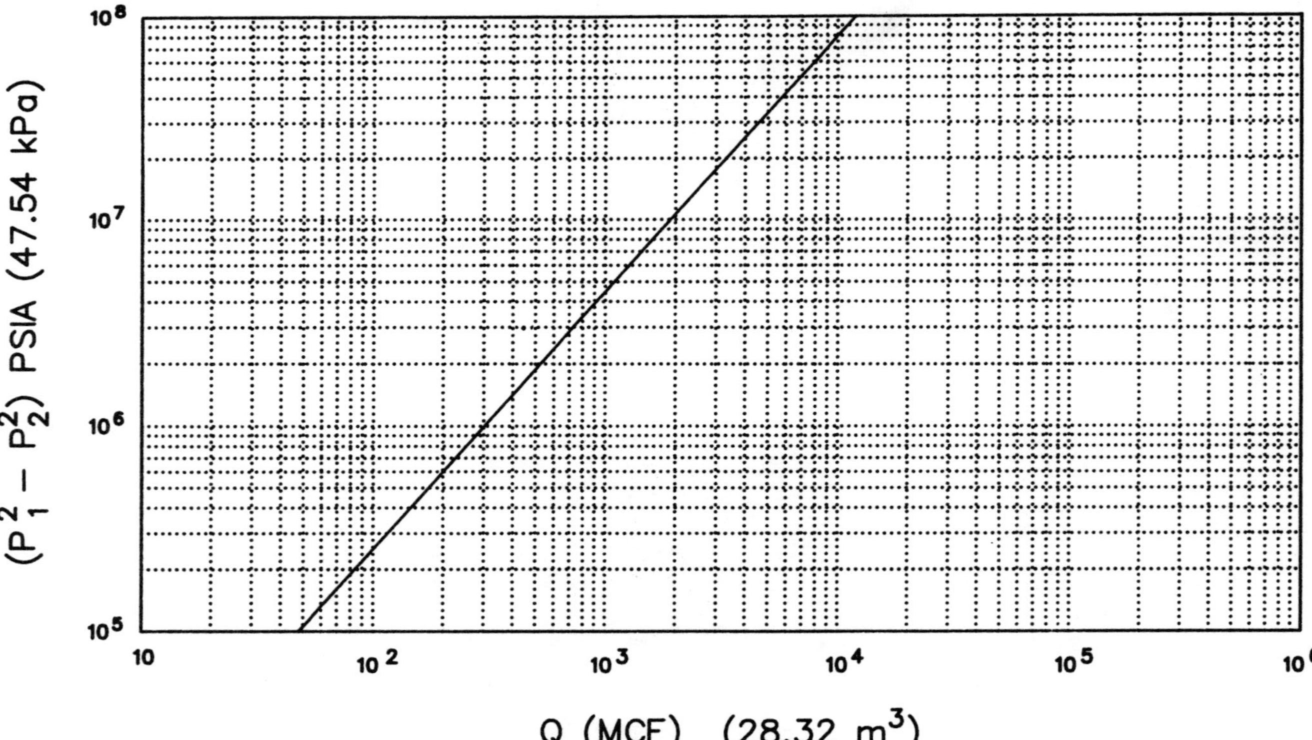

Figure 37. Wharton deliverability curves—Area 1.

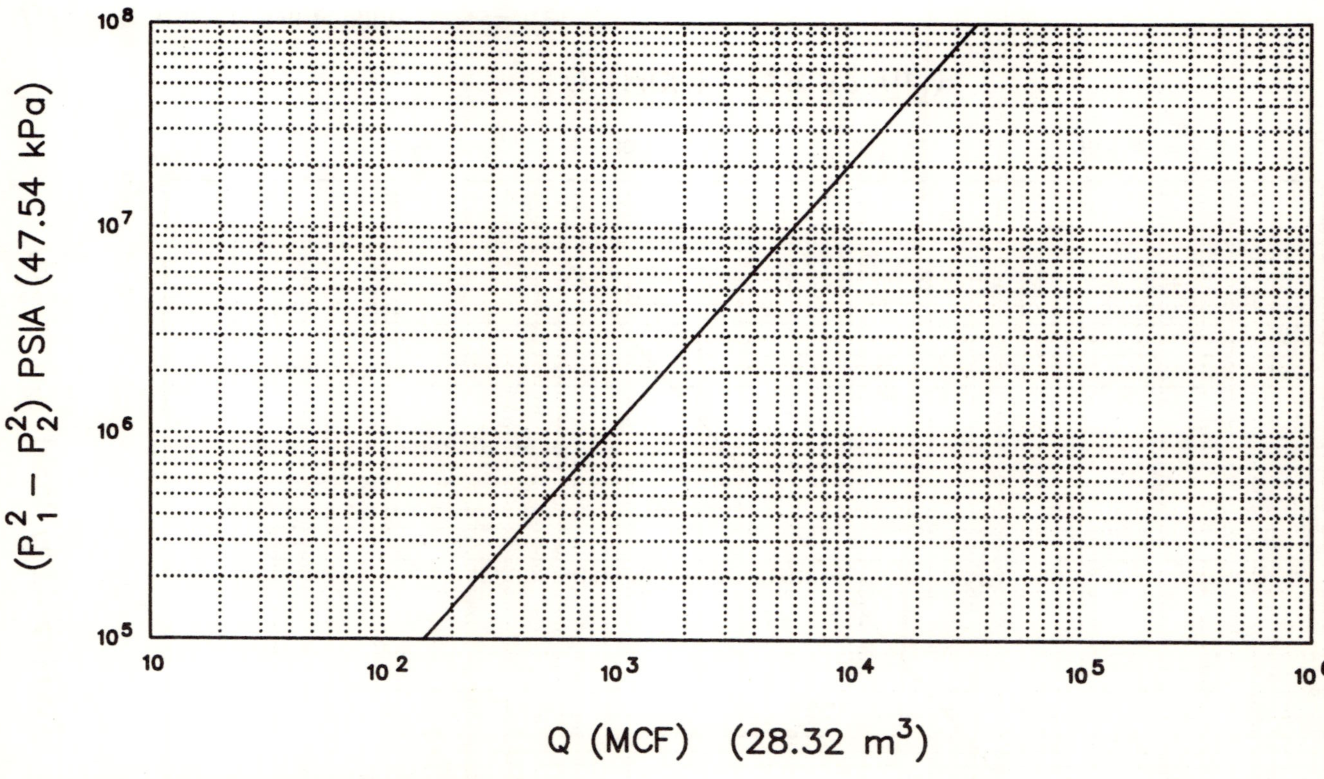

Figure 38. Wharton deliverability curves—Area 2.

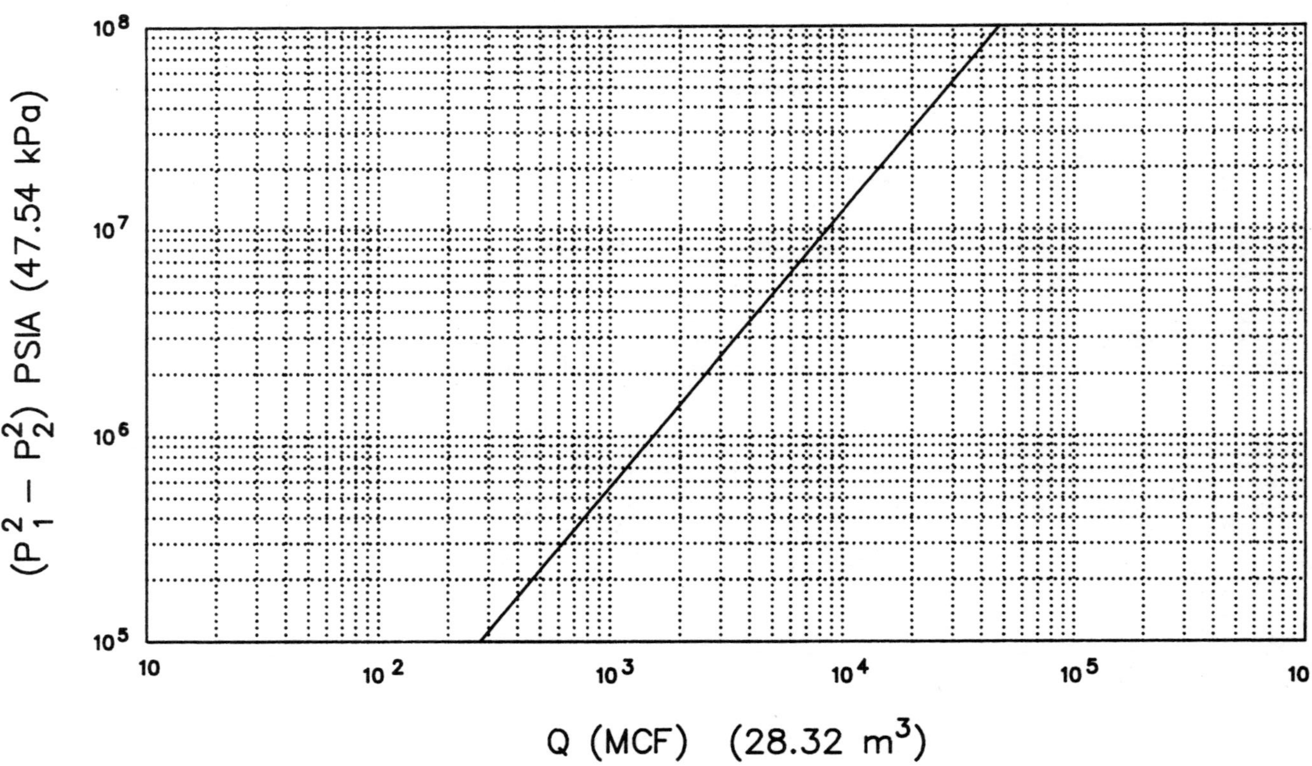

Figure 39. Wharton deliverability curves—Area 3.

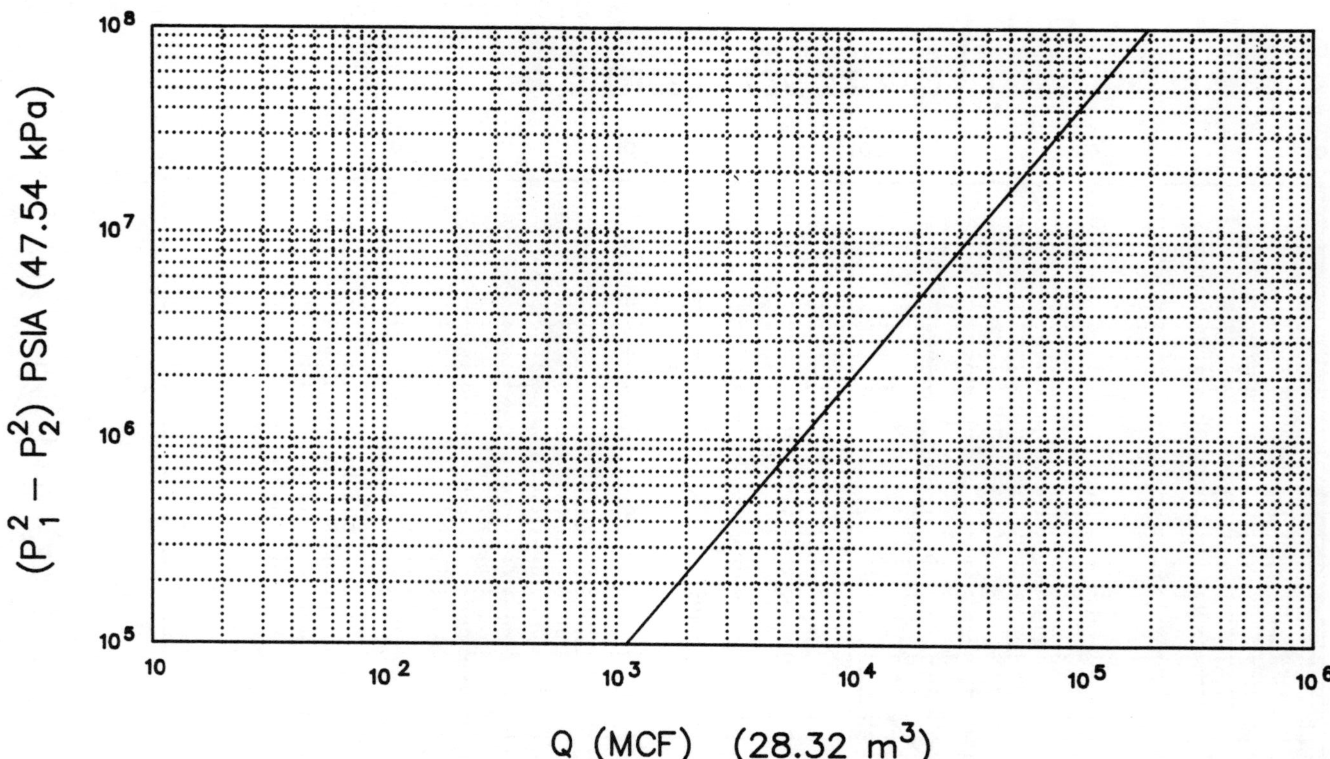

Figure 40. Wharton deliverability curves—Area 4.

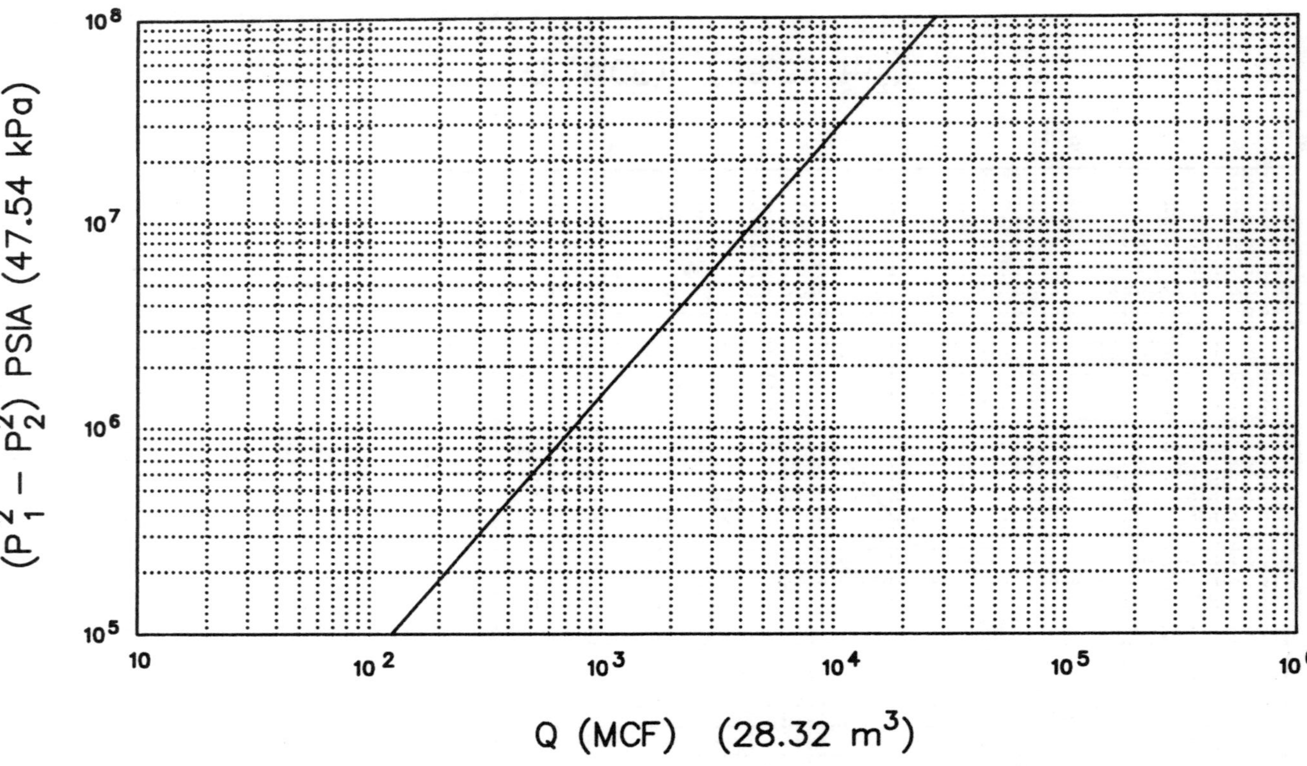

Figure 41. Wharton deliverability curves—total field.

The East Fork area of the reservoir has never been developed or used for the injection and withdrawal of gas. Deliverability there was so low that it was economically impossible to develop. In the First Fork area, the deliverability is also very poor and can be only partially used in the operation of the storage field.

Core data and logging data taken from wells located in the main part of the reservoir indicated permeability of 3.5 to 5.0 mD in the upper 7 to 10 ft (2.1 to 3.0 m) of sand. The remaining 20 to 30 ft (6.1 to 7.0 m) of Oriskany sandstone indicated a matrix permeability of less than 0.1 mD. Most of the deliverability from this field is derived from natural fractures that occur close to major fault lines and induced fractures. Without the naturally occurring fractures in the Oriskany sandstone and the induced fracturing of all the wells, this field probably would not have been developed for storage. Even with the low permeability, the company has, through special operating techniques and fracturing, used this area for the storage of gas in a cost effective manner.

WELL DESIGN AND COMPLETIONS

In this old production field, three basic types of wells were to be developed, calling for various designs and completion techniques. They were as follows:

- Plugged and abandoned wells requiring reconditioning for storage (see Fig. 42 for a sketch).
- Existing production wells requiring reconditioning for storage. (For sketches of cased and open-hole recompletions see Figs. 43 and 44.)
- New wells drilled for storage (see Fig. 45).

These wells were all completed basically the same way, in that the objective was to have a new string of casing—either 4½-in. (114.3-mm)-OD, N-80, 11.6 lb/ft (17.26 kg/m); or 7-in. (177.8-mm)-OD, N-80, 26 lb/ft (38.69 kg/m)—set through the storage horizon and selectively perforated. All long strings and intermediate strings were cemented from the bottom to the surface.

The reworking of a plugged hole was found to be the most difficult to recondition and replug because of the original construction technique of cutting or ripping the production string just above the cement used to hold the casing in the hole. Because of this technique, operators had to clean out down to the stub of casing left in the hole and then get inside the stub, clean it out, and drill through the Oriskany. A new string of smaller casing was then run through the stub and into the Oriskany formation and cemented to the surface. If the well had to be replugged, the technique of cleaning out into the Oriskany and filling the hole from bottom to the top with cement was still followed. All plugged holes were reworked with a rotary rig.

An active well is defined as a well that remains as it was when originally drilled for production. The active wells were probably the easiest and least costly of all the wells to recondition and plug. The reconditioning involved cleaning the well out to the bottom through the 7-in. (2.1-m) string of casing and running a cement bond log on the original production string. If good bond was found on the bottom 500 ft (152.4 m) of the production string, efforts were made to drill through the 50 ft (15.2 m) below the Oriskany. If the cement bond was not good, the old string would be perforated and cement circulated behind the old string of casing. Then a new, smaller casing [4½ in. (114.3 mm)] was run and set below the Oriskany and the well was cemented from the bottom to the surface in two stages. If the well was to be plugged, it would be cleaned out into the Oriskany, the original production string would be pulled and the entire hole would be pumped full of cement. All active wells were reconditioned or plugged with a cable tool rig.

All new wells drilled in the field were drilled for a 7-in. (177.8-mm)-OD production string set through the Oriskany and cemented from bottom to surface. All new wells were drilled with a rotary rig.

CASING EQUIPMENT AND CEMENTING MATERIALS

The casing equipment and cementing materials used to complete the wells in the storage field consisted of the following:
- Float shoe
- Float collar
- Two-stage cementing collar
- Two cement baskets
- Twenty casing centralizers
- Bottom four joints coated with Halliburton resin and sand
- Unvarnished casing

The float shoe and float collar were used to ensure that the cement slurry remained static while the cement was setting and also to contain the cement behind the casing.

A two-stage cementing procedure was employed to prevent the possibility of a premature fracture of shale formations during the cementing operation. It was also used to prevent dehydration of the cement resulting from high pressures involved in circulating the system.

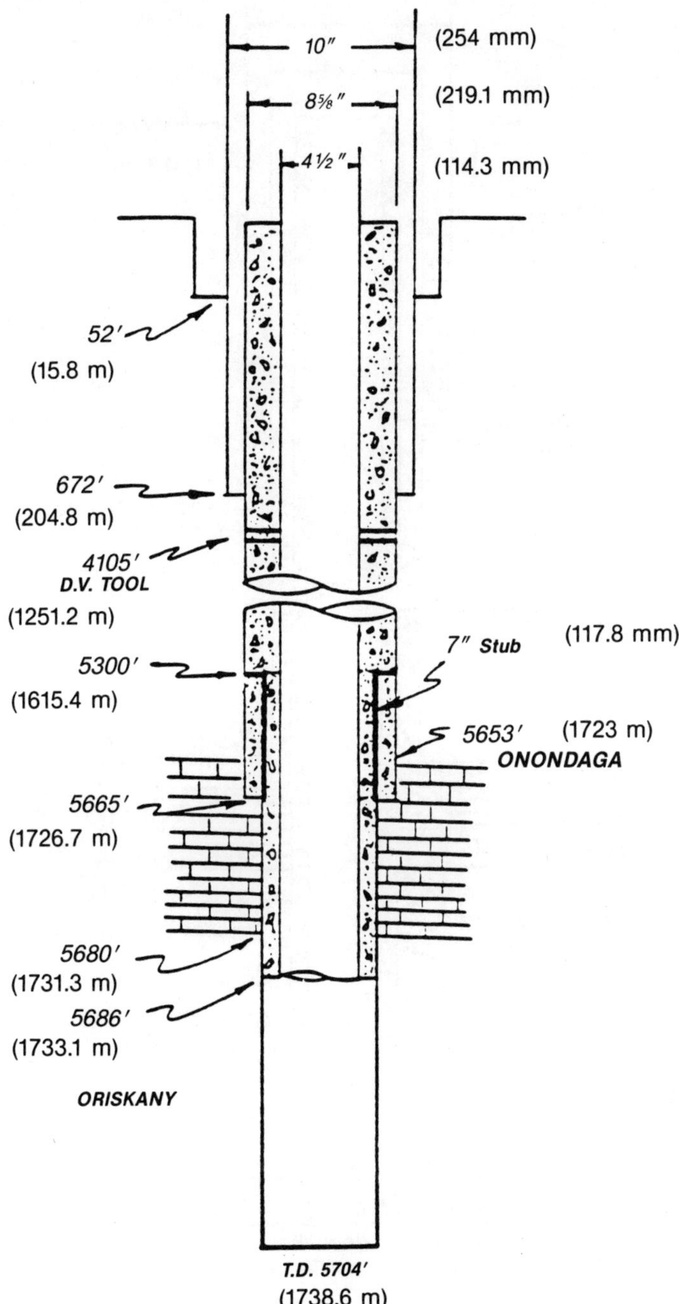

Figure 42. Plugged and abandoned well recompleted for storage.

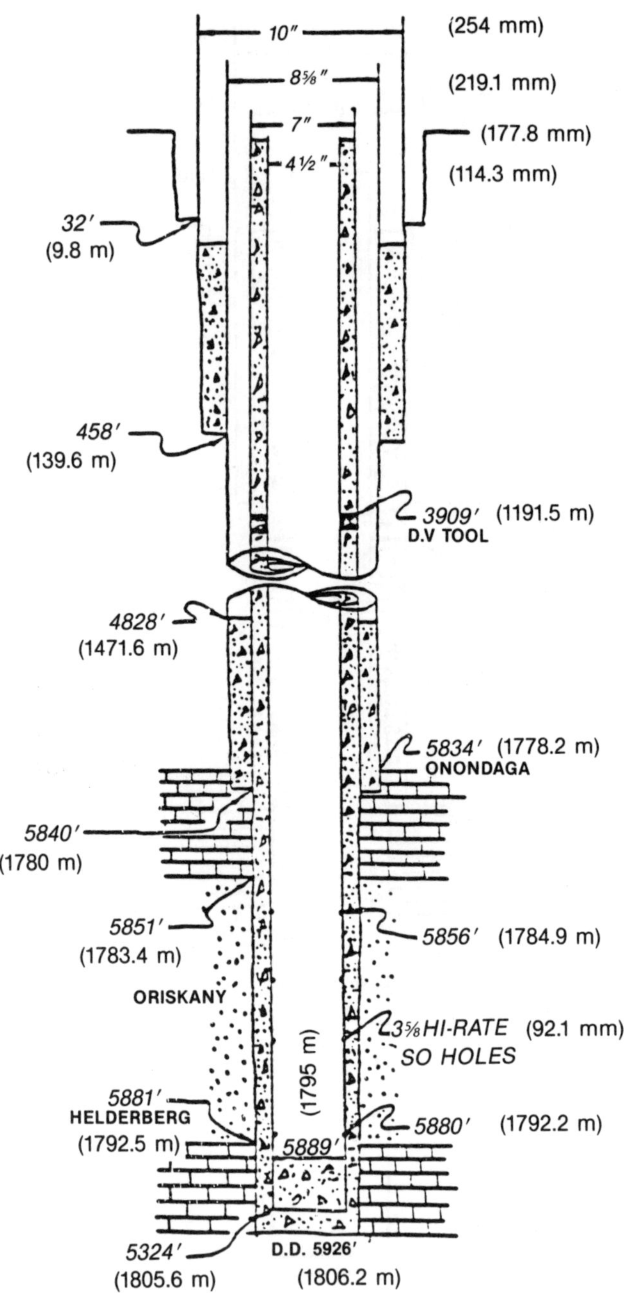

Figure 43. **Existing cased production well recompleted for storage.**

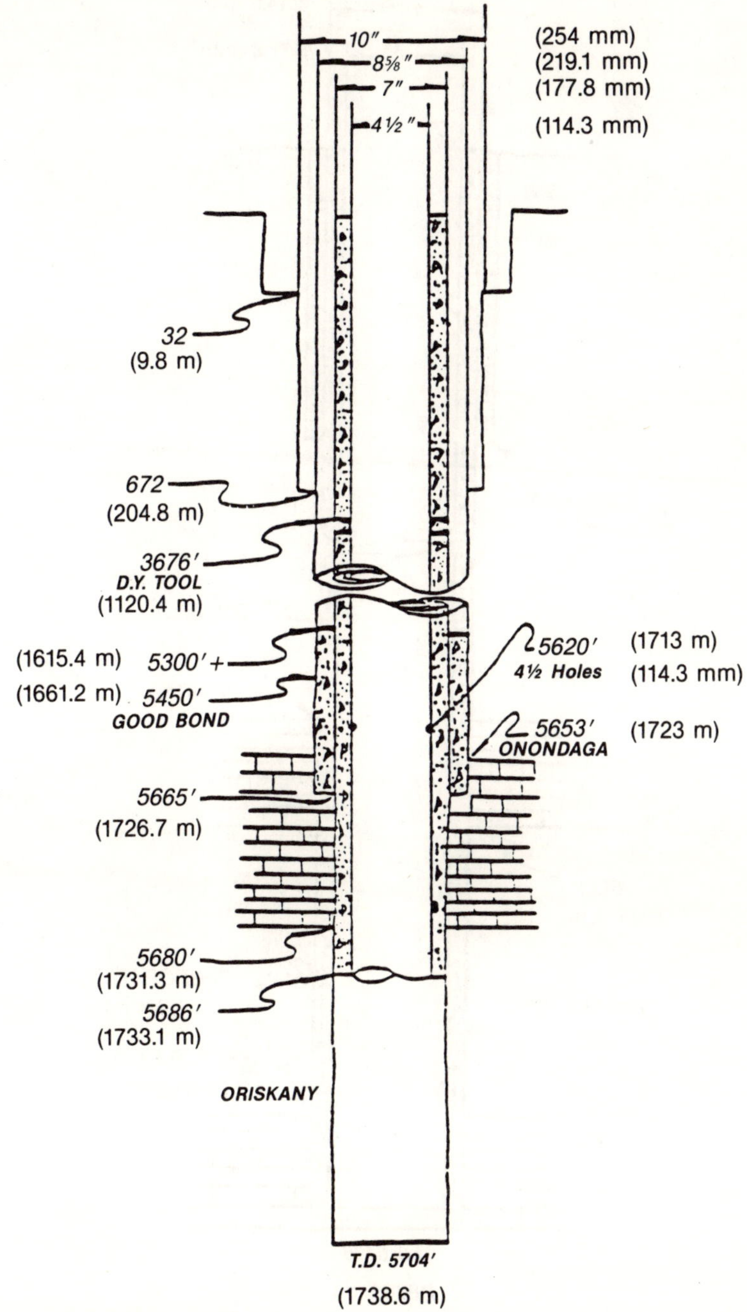

Figure 44. **Existing open-hole production well recompleted for storage.**

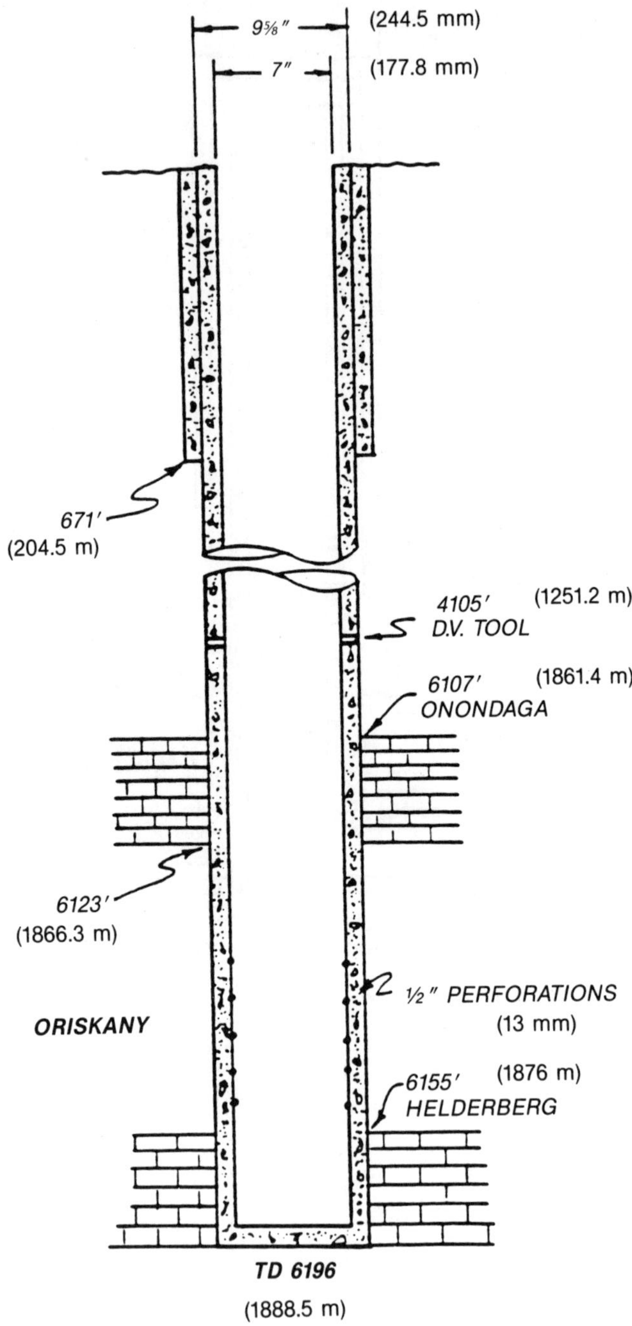

Figure 45. New wells drilled for storage.

The two-stage cementing tool was placed 2 000 ft (609.6 m) up from the bottom with two cement baskets placed on the second and third joints below the stage tool. The cement baskets helped to relieve the hydrostatic head of the cement column on the formations.

The 20 casing centralizers were placed on the flow string as follows: one each on the bottom four joints, one each above and below the two-stage cementing tool, and 14 spaced evenly throughout the remaining portion of the flow string.

The bottom four joints were coated with resin and sand to secure a better cement bond. The entire casing string was delivered unvarnished to further ensure a good cement bond.

The cement used on flow strings was as follows: in the first (bottom) stage, a 50–50 Pozmix cement (29 percent gel, 18 percent salt, 0.8 percent Halad-9) with a slurry weight of 14.5 lb/gal (1.74 kg/L) was used. In the second stage, a 50–50 Pozmix cement (2 percent gel) with a slurry weight of 14.1 lb/gal (1.69 kg/L) was used. The safety factor on all volumes was 25 percent.

On the rotary holes where mud was used as a circulating medium, a mud clean-out agent was run ahead of all cement. The shallow string of 8⅝-in. (219.1-mm) or 9–5/8-in. (244.5-mm) casing was cemented with a 50–50 surface Pozmix, 2 percent calcium chloride, ¼-lb (0.11-kg) per sack of flocele. Attempts were made to cement from the bottom of the casing to the face in all cases.

LOGGING PROGRAM

Most of the logs run during the development of the storage field were used basically for completion purposes. However, several reservoir evaluation-type logs were run on a cross section of wells in the field. The logs most generally used for completion purposes were gamma-ray and caliper, temperature, neutron, and tracer.

PERFORATING

As stated previously, the main objective was to complete all wells through the storage horizon and perforate. Because many of the original production wells in this field had been shot with nitroglycerin, there was a large hole in the Oriskany formation, making it impossible to complete wells of this type by setting through and perforating. In these wells the casing string was set in the top 3 ft (0.9 m) of the Oriskany and cemented back to the surface. In wells where setting the flow string through the Oriskany was successful, the company perforated using 3⅝-in. (92.1-mm) Hi-Rate Jets, two per foot (meter) covering a 24-ft (7.3-m) section of the sand. The objective was to create a good, round perforation in the casing and eliminate any possibility of rupturing or splitting the pipe and damaging the cement bond. The 3⅝-in. (92.1-mm) Hi-Rates were the proper strength and made the proper diameter hole for deliverability and completion of operating wells in the field. Attempts were also made to perforate the wells within 3 to 4 days after cementing the wells. The procedure appears to have been satisfactory, because no failures were experienced during the fracturing of the wells.

STIMULATION

After a well was completed through the sand and perforated, the well was acidized with 1 000 gal (3 785 L) of 10 percent hydrochloric acid. The acid was pumped into the formation very slowly to allow time for it to clean out the perforations. In a well completed open hole, acidizing was not performed until the well was to be fractured; in both cases, the frac job would be spearheaded with 1 000 gallons of 10-percent mud clean-out agent and/or 10-percent hydrochloric acid.

The hydraulic fracturing program used to stimulate all of the operating wells in the storage pool consisted of a 15 000-gal (56 781-L) water frac with 2-percent calcium chloride. The water was gelled with 650 lb to 800 lb (295 kg to 363 kg) of Halliburton WG-5 3500. Proppant consisted of 100 to 200 sacks of 20- to 40-mesh sand tailed in with 20 to 25 sacks of 12- to 20-mesh sand.

The treating pressure was generally between 5 000 and 5 500 psig (34 475 and 37 922 kPa) and resulted in an injection rate of approximately 25 bbl (3 974 L)/min. The breakdown pressure in most cases was very high because of the low permeability of the reservoir sand. The instant shut-in pressure was generally around 4 300 psig (29 648 kPa).

SURFACE FACILITIES

The original design of the storage areas as a gas storage pool was based on and is being operated at a maximum stabilized wellhead pressure of 3 600 psig (24 822 kPa). The gathering system, trunk lines, and all compressor station piping and engines were therefore designed for a common injection and withdrawal system to accommodate an operating pressure of 3 600 psig.

The well lines, gathering system, and trunk lines were also designed to flow at least 300 MMcf (8.5×10^6 m³)/d on withdrawal and 118 MMcf (3.3×10^6 m³)/d on injection. The entire system was not to exceed a 25-psig (172-kPa) loss from any point on the system to the compressor station. The compressor station was designed on a minimum of 800-psig (5 516-kPa) suction pressure and a 1 200-psig (8 274-kPa) discharge pressure on withdrawal. The station also had to be capable of injecting gas at a suction pressure of 1 000 psig (6 895 kPa) and discharging at 3 800 psig (26 201 kPa). Calculations with several combinations of horsepower (kilowatt) and wells for both injection and withdrawal were made. The results of these calculations to meet the above conditions were 88 wells and 8 300 horsepower (6 189 kW).

OPERATIONS

Because of the low and different permeabilities in this reservoir, the operational procedures are somewhat different than most storage fields.

During the injection of gas, all wells in areas 1 and 2 (see Fig. 36) are put on line first and areas 3 and 4 are put on line at approximately mid-season or when the injection pressure reaches approximately 2 800 psig (19 306 kPa). The reason for this pattern of injection is to optimize the volume of injected gas into the low-permeability areas when the pressure in the reservoir is low.

The procedure on withdrawal is the same as that on injection: area 1 and 2 wells are turned on line first and, as additional rates are required, area 3 is turned on. Area 4 wells are used primarily for peaking situations. Also on extreme peaking conditions, all indicator wells can be used for the withdrawal of gas. The main reason for not withdrawing gas from the high-deliverability areas in the main pool early in the season or during the season is to avoid creating low-pressure areas. If pressure sinks developed in the high-permeability portion of the field, less gas could be withdrawn from the reservoir and the deliverability rate would suffer.

It should also be pointed out that all wells in the Wharton Field are turned on and off individually for both injection and withdrawal; that is, reservoir pressure is not floated up against a main line valve to the compressor station, allowing all wells to feed simultaneously.

PRESSURE TESTS

At the conclusion of the injection cycle and withdrawal cycle, a 7-day shut-in test is conducted. Deadweighted pressure tests are taken 24 hours after starting the shut-in test, and gauge pressures are taken for the remainder of the test until the last day, when deadweight pressures are again taken. These data are used to evaluate inventories and to establish stabilized reservoir pressures throughout the entire area of the field.

WELL MAINTENANCE

Throughout the year, attempts are made to take deliverability and injectability tests on each well. As part of the well maintenance program, gamma-ray, neutron, and temperature logs are run on 20 percent of the wells each year.

Chapter 7

DEPLETED HIGH-PERMEABILITY OIL RESERVOIR

Aliso Canyon
Southern California Gas Company
George E. Babashoff, Larry D. Krohmer, William A. McEachin,
Marvin E. Melton and Ali Razavi

INTRODUCTON

For a case study of a depleted high-permeability oil reservoir, this chapter describes Aliso Canyon storage field, an important facility of the gas utility system in southern California. The storage field was acquired in the early 1970's and developed to its present capacity over the succeeding decade. This chapter covers the engineering and operating practices that are pertinent to the acquisition, development, and operation of Aliso Canyon.

PRELIMINARY PLANNING

The need for underground storage of natural gas to provide reliable service to gas customers has existed in southern California for more than 40 years. The first storage field was developed in 1941 at Goleta near Santa Barbara (see Fig. 46 for a map showing major pipelines and storage fields in the system). Three other storage fields near Los Angeles were developed in the next two decades as the natural gas demand increased in proportion to the large population growth in southern California.

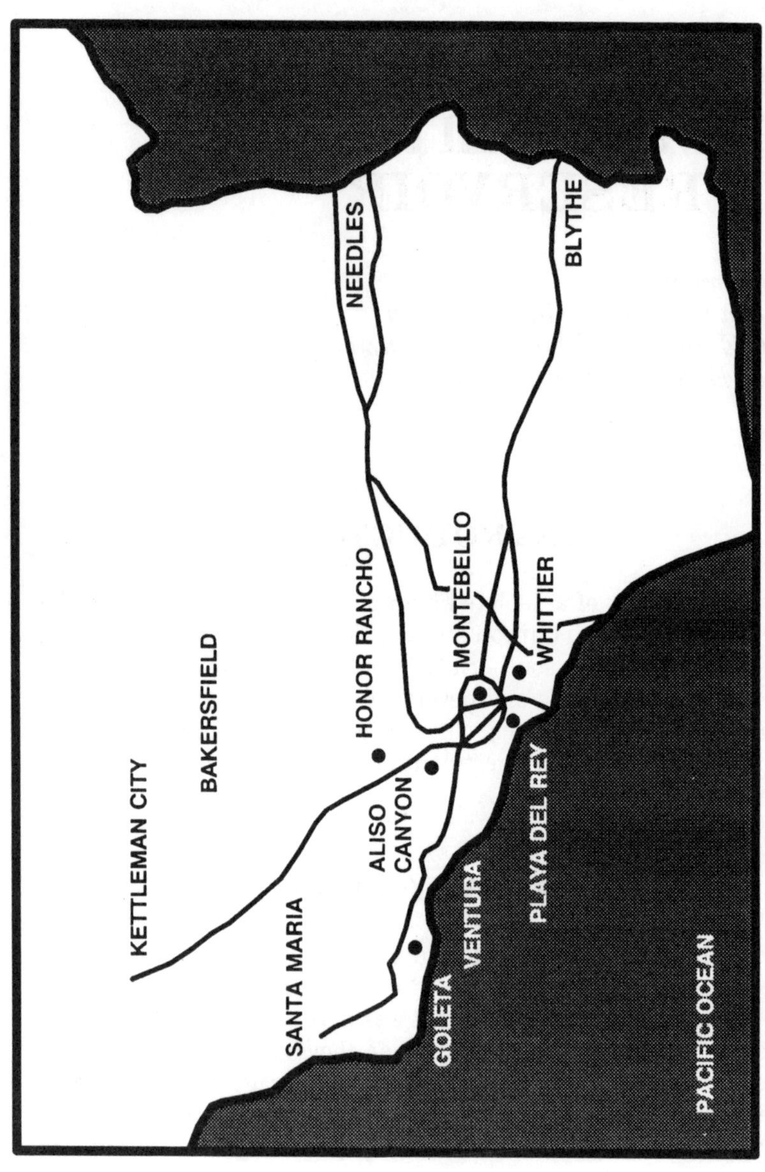

Figure 46. Major pipelines and underground storage fields in the Pacific Lighting Utility System, southern California.

By the early 1970s there were already more than 3 million gas customers in southern California, and the prospect was strong for comparable continued growth. However, at the same time gas supplies were diminishing rapidly and regulatory restrictions on the end uses of natural gas were increasing. These coinciding events caused the conversion in 1972 of a very large semidepleted oil and gas field, named Aliso Canyon, to underground storage.

This fifth system storage field, located north of Los Angeles proper, was chosen because it was capable of providing very large peaking and seasonal load equation volumes. The initial plan included a peak-day capacity of 1 Bcf/d (28.3×10^6 m³/d) and an annual cycle capacity of 70 Bcf (1.98×10^9 m³). Injection capacity was to range from 500 MMcf/d (14.16×10^6 m³/d) at low storage inventories to 300 MMcf/d (8.50×10^6 m³/d) at higher inventories. All facilities to obtain these objectives were to be available within 3 years. In subsequent years the peaking capacity was raised to 1.5 Bcf/d (42.48×10^6 m³/d), and there is recognized potential to expand the peak deliverability to 2.0 Bcf/d (56.63×10^6 m³/d) if necessary in the future. This major storage field represents 60 percent of the total working storage capacity in six storage fields presently located in the southern California utility system.

GEOLOGY

The Aliso Canyon storage field lies within the Transverse Range Geologic Province of southern California, a highly faulted, structurally complex province, with east-west trending mountain ranges bounded by major thrust and strike-slip faults (see Fig. 47). It cuts obliquely across the more extensive northwest-southeast-trending San Andreas fault system. The Aliso Canyon area contains a sequence of 17 000 ft (5 182 m) of Cretaceous Age through Pleistocene Age sedimentary rocks, extensively deformed by north-south shortening between two rigid blocks, the Simi Hills to the southwest and the San Gabriel Mountains to the northeast. The dominant structural element in the Aliso Canyon area is the Santa Susana Thrust fault, which delineates the southern margin of a deep-water depositional trough of folded sedimentary rocks of Middle Miocene to Plio-Pleistocene age. Fig. 47 shows the location of the Santa Susana fault and the Aliso Canyon area with respect to onshore basins and faults in southern California.

The Aliso Canyon Field is located in the East Ventura Sedimentary Basin in Los Angeles County, approximately 25 mi (40 km) northwest of central Los Angeles. Underlying the south slope of the Santa Susana Mountains, it is one of the most structurally complex oil and gas fields

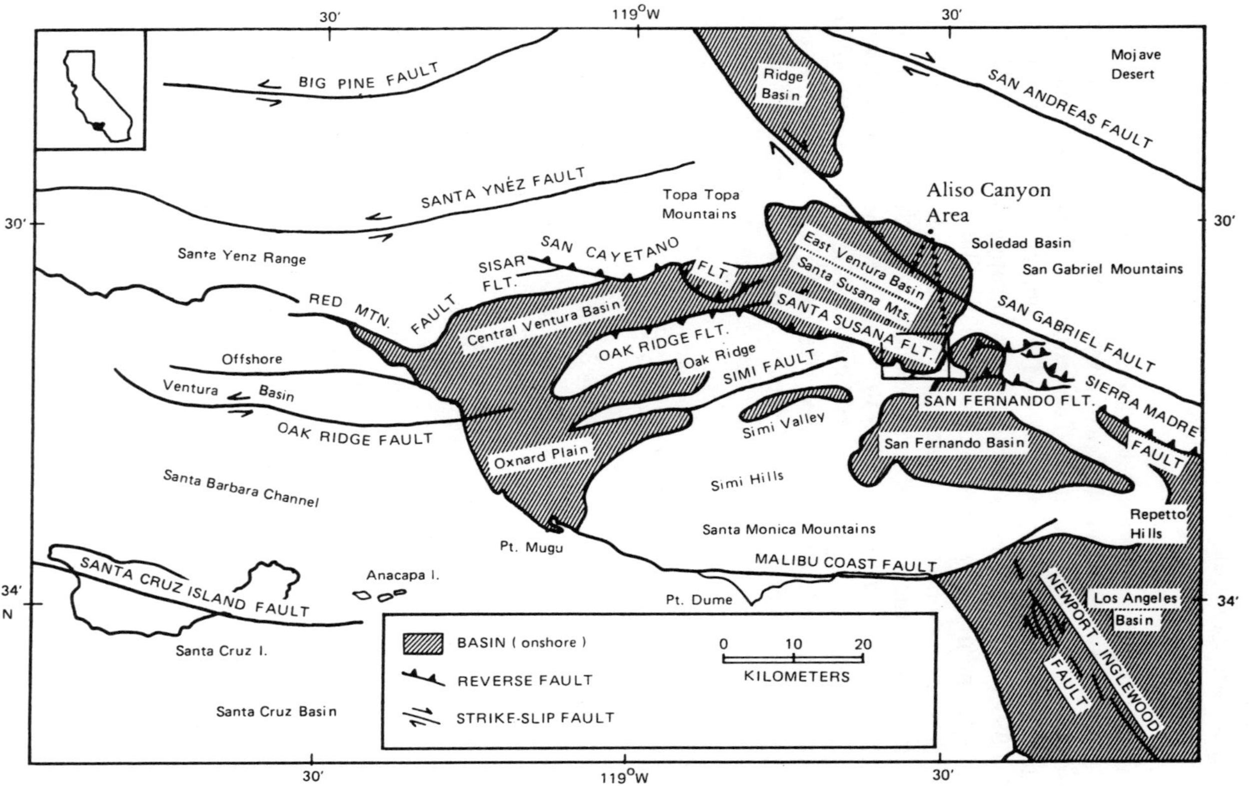

Figure 47. **Location of Aliso Canyon area in relation to major onshore basins and faults.**

in California. Aliso Canyon field is basically a faulted, plunging anticline with a complicated set of structural and stratigraphic conditions accounting for the original hydrocarbon accumulations.

The field was discovered in 1938 with the Tidewater Associated Oil Company's drilling of Porter No. 1 well. The original well produced from a shallow oil zone named the Porter zone, but later field development resulted in the discovery of four additional zones. Oil production has come from Pliocene, Miocene, and Eocene age sediments. The five producing zones in order of increasing depth are named Aliso, Porter, Del Aliso, Sesnon, and Frew. Field development continued for several years at a generally steady pace until 1955, when approximately 120 wells were producing.

STRUCTURE

The Aliso Canyon oil field is a southeasterly plunging anticline with structural closure on the south and east (see Fig. 48 for a contour map). Closure on the north is provided by the extensive Santa Susana Thrust fault, the Ward fault, and, to a very small extent, by the Roosa fault. Western closure is effected by the Frew fault.

The Santa Susana is a north-dipping fault system or zone with a vertical or overthrust displacement of about 5 000 ft (1 524 m) and a horizontal or strike-slip displacement of from 2 to 3 mi (3.2 to 4.8 km). There is a significant difference in geology above and below the fault zone; Miocene strata are thrustover Pliocene sediments (see Fig. 49 for a cross-section map). The fault zone dips about 15° to the north; however, that dip increases rapidly to about 80° near the northern boundary of the field.

The Ward fault, which provides the northern closure on the storage zone, is a reverse fault with displacement varying between 400 and 800 ft (122 and 244 m). Bringing Miocene rocks over Pliocene age sediments, it strikes east-west and dips 60° south.

The Roosa fault dips to the north, displacing the storage zone between 800 and 1 300 ft (244 to 396 m). This late Pliocene fault merges with and is truncated by the younger Santa Susana fault of Pleistocene age.

The Frew fault, near the western field boundary, is a high-angle reverse fault with a vertical displacement of more than 5 000 ft (1 524 m). It comprises numerous fault plates, the fault zone ranging from 300 to 900 ft (91 to 274 m) wide. This fault dips about 60° to the southwest, bringing Cretaceous, Paleocene, Eocene, and Pliocene age rocks over those of the Upper Pliocene age. The Frew fault steepens to 75° as it approaches the intersection with the younger Santa Susana fault.

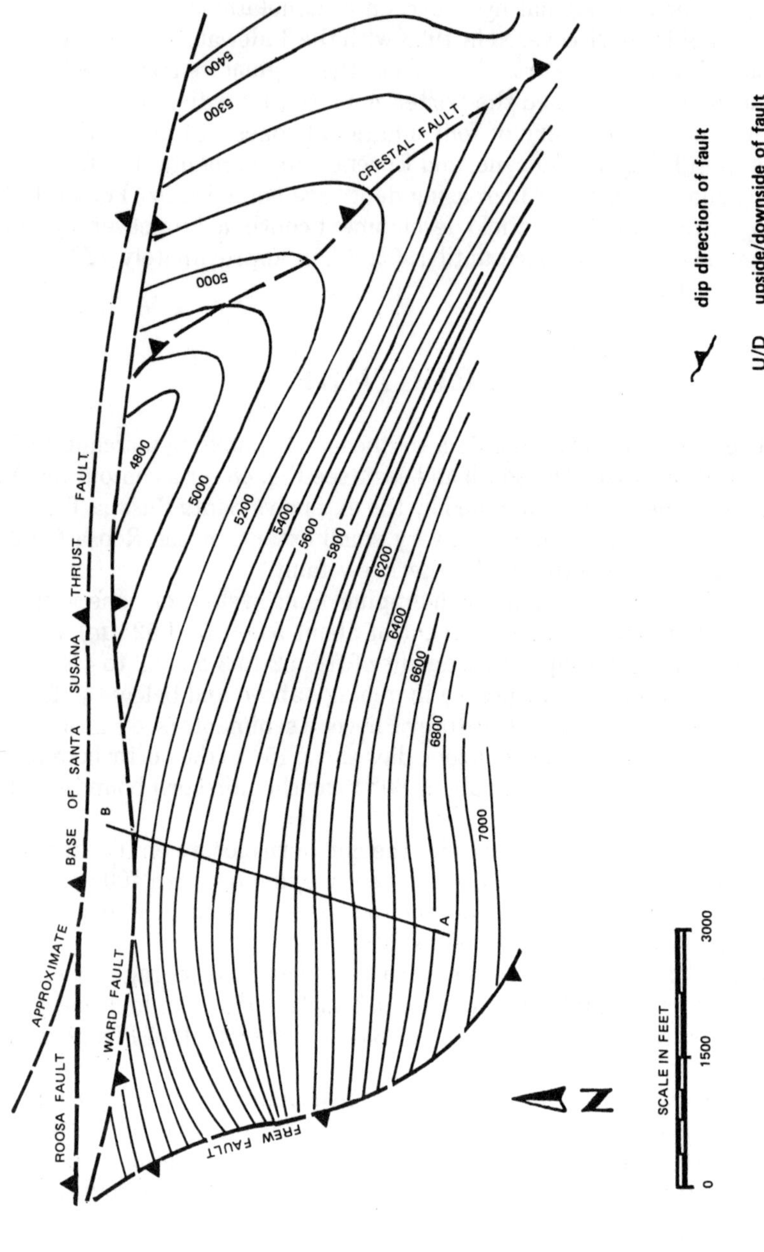

Figure 48. Contour map of Aliso Canyon oil field.

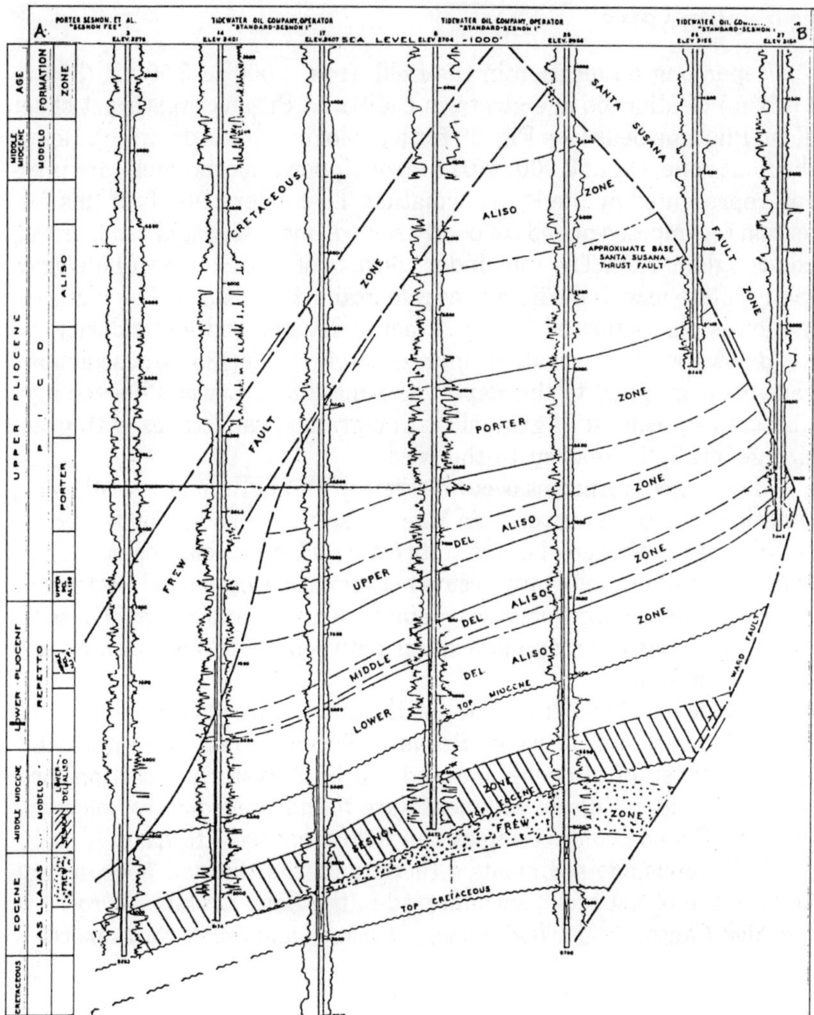

Figure 49. Cross-section map of Aliso Canyon oil field.

The Crestal fault trends along the Miocene-Pliocene structural axis suggesting the initial formation of the Aliso Canyon structure during Lower Pliocene time. It dips to the east with approximately 100 ft (30 m) of displacement.

There are several small normal faults in the field with typical displacements less than 75 ft (23 m). A few of these faults may affect pressure distribution within the field, but most seem to allow migration of gas through sand to sand contacts.

STRATIGRAPHY

Depending on location in the field, from 1 000 to 5 000 ft (305 to 1 524 m) of Miocene age strata in the Santa Susana overthrust sheet lie on Pliocene beds (see Fig. 50 for a geologic column drawing). Below the fault zone, about 3 000 ft (914 m) of Upper Pliocene rocks are present, represented by the Pico Formation. The upper 1 000 ft of this formation is composed of beds of poorly sorted, unconsolidated sand, gravel, and conglomerate. The remainder consists of thick zones of blue-gray, sandy siltstone with medium to coarse-grained sandstone. The Pico Formation is characterized by rapid facies changes and lenticularity.

Below the Pico Formation, approximately 600 ft (183 m) of sediments have been assigned to the Repetto Formation of Lower Pliocene age. These strata consist of gray, silty, fine-grained sandstones with grain size generally increasing to the west.

Major unconformities occur at the top and bottom of the Miocene, resulting in a comparatively thin Miocene section of about 650 ft (198 m). In some areas Miocene beds rest directly on Cretaceous rocks. A third major angular unconformity occurs between the Eocene and Cretaceous horizons. These unconformities result in sand-to-sand contacts in some areas, allowing intercommunication between the Frew, Sesnon, and Cretaceous layers.

As much as 1 000 ft (305 m) of Eocene sediments has been found in some Aliso Canyon wells. Because Eocene strata unconformably underlie those of Miocene age and dip in a westerly direction, progressively older Eocene sediments are found immediately below the Miocene-Eocene contact across the field from west to east.

Cretaceous age sediments directly underlie those of Miocene age and are the oldest rocks encountered in the field. Cretaceous rocks in the Aliso Canyon field are siltstone, sandstone, and cobble conglomerates.

STORAGE ZONES

Sesnon Zone

The primary gas storage zone in the field is the Sesnon sands. Directly overlying these sands are 150 to 350 ft (46 to 107 m) of Middle Miocene Modelo shales with interbedded siltstones. This Miocene shale is impermeable to gas and accounts for the very large gas cap that existed at the time of discovery. The original contact area of shale caprock with the gas cap exceeded 500 acres (202 ha). The underlying Sesnon sands are Middle Miocene age, ranging in thickness from 200 to 400 ft (61 to 122 m). Effective sand thickness is only about 110 ft (34 m).

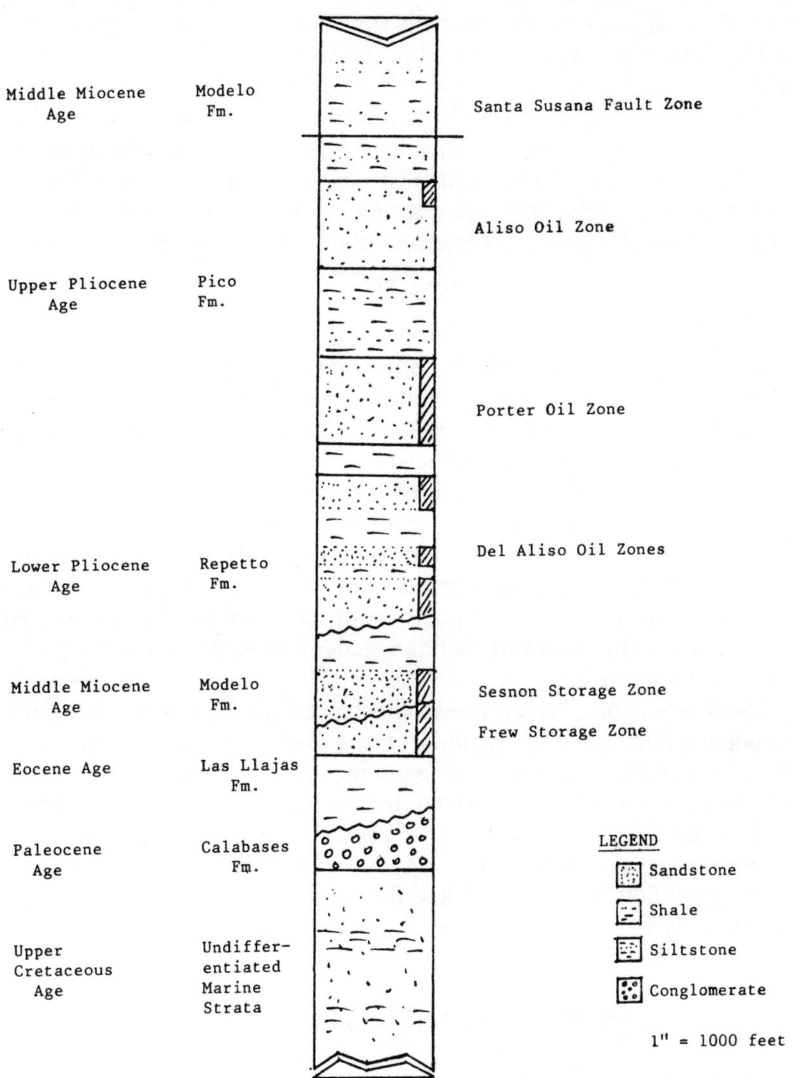

Figure 50. General geologic column of Aliso Canyon storage field.

The Upper Sesnon is made up of shallow Miocene bar sands deposited on the offshore edge of the Miocene shelf. They were later modified by north-trending channel deposits. Core studies indicate that these bar sands were deposited in water depths of 50 to 200 ft (15 to 61 m). Individual sands average 50 ft (15 m) thick and the total thickness of the Upper Sesnon averages 150 ft (46 m). There was east-west development of these sands with apparent thinning to the north and south.

The Lower Sesnon unit consists of 50 to 300 ft (15 to 91 m) of interbedded sandstone, siltstone, and shale with a basal member containing lenses of conglomerate. It is a fluvial and near-shore marine sequence representing the initial stage of Miocene transgression in this area. The Lower Sesnon lies with angular discordance on older Eocene and Cretaceous units. Individual sands can range up to 110 ft (34 m) in gross thickness and are generally aligned east to west.

Frew Zone

The Frew zone is a thick conglomerate and sandstone wedge underlying the Sesnon zone. Constituting the basal reservoir of the gas storage field, these sands are believed to represent the development of a fluvial system on the post-Cretaceous erosional surface. The zone has limited areal extent and is confined primarily to the western portion of the field. It is found at a depth of 8 650 to 9 075 ft (2 637 to 2 766 m). In many areas, intercommunication of gas occurs through sand-to-sand contact of the Frew and overlying Sesnon sands. In effect, these zones act as one reservoir. In other portions of the field, the Frew is sealed by up to 300 ft (91 m) of interbedded shales and siltstones of the Santa Susana formation.

The Frew sands, which are in the Las Llajas Formation of Upper Eocene age, unconformably underlie the Miocene. Post-Eocene uplift and erosion have resulted in an incomplete Eocene section across the field, and the Frew zone is missing in many wells. Pinchout of the Frew sands occurs from northeast to southwest. South of the Frew pinchout, the Sesnon zone overlies the Cretaceous sediments. The Frew zone ranges up to 500 ft (152 m) thick and consists of coarse-grained, friable sandstone with occasional igneous pebbles and interbedded shale.

LAND AND RIGHT-OF-WAY

In 1972 a major task force was organized to acquire the land, leases, permits, and rights-of-way; design the facility; and supervise the installation of all the equipment required to operate the field. The original objective of the task force was to have 70 Bcf (1.98×10^9 m³) of working inventory in place on November 1, 1974, and a cycle capacity availability of 70 Bcf (1.98×10^9 m³) per year beginning May 1, 1975.

The Aliso Canyon field was operated as a multizone multioperator, oil and gas field before the acquisition of the Sesnon and Frew zones for gas storage. These same operators were, and presently are, producing oil and gas from horizons above the storage zone.

The surface above the entire storage zone of 1 970 acres (797 ha) and a buffer zone of 1 230 acres (498 ha) surrounding the storage area were acquired to a depth of 500 ft (152 m) (land boundaries are shown in Fig. 51). The purpose of the buffer zone was to isolate the storage field by sight and sound barriers from adjacent residential areas.

As stated earlier, the gas storage reservoir consists of the Sesnon and Frew zones. Oil and gas leases, mineral rights, fee interests, working interests, royalty and overriding royalty interests, and the indigenous natural gas in the Sesnon and Frew zones, were all acquired by the storage field operator.

Wells and associated equipment serving the storage zone and required in the storage operation were also acquired by the storage field operator. In addition, the rights to inject and remove gas stored in the Sesnon and Frew zones were acquired, as well as the right to recover storage gas that might have migrated to zones above the storage zone. The ability to monitor gas production from wells completed above the storage zone was negotiated with other field operators, who were continuing to produce oil zones above the storage reservoir. These rights were acquired from all of the owners of such interests except the right of the federal government in the federal oil and gas leases.

The Sesnon and Frew zones are located between 6 812 ft (2 076 m) and 9 354 ft (2 851 m) below the surface of the ground. Therefore, the landowners and operators reserved the right to drill through the storage zone to depths below the storage zone. However, the reservation provided that the drilling operation must be approved and supervised by the storage field operator

REGULATORY REQUIREMENTS

In 1968 the Sesnon and Frew zones were unitized (including federal lands) to conduct a waterflood project. The unit agreement, which contained provisions for gas storage in the unitized zones, was approved by the California Division of Oil and Gas and the Department of the Interior. The operators entered into a gas storage enabling agreement for the storage of natural gas at this time, which was also approved by state and federal agencies.

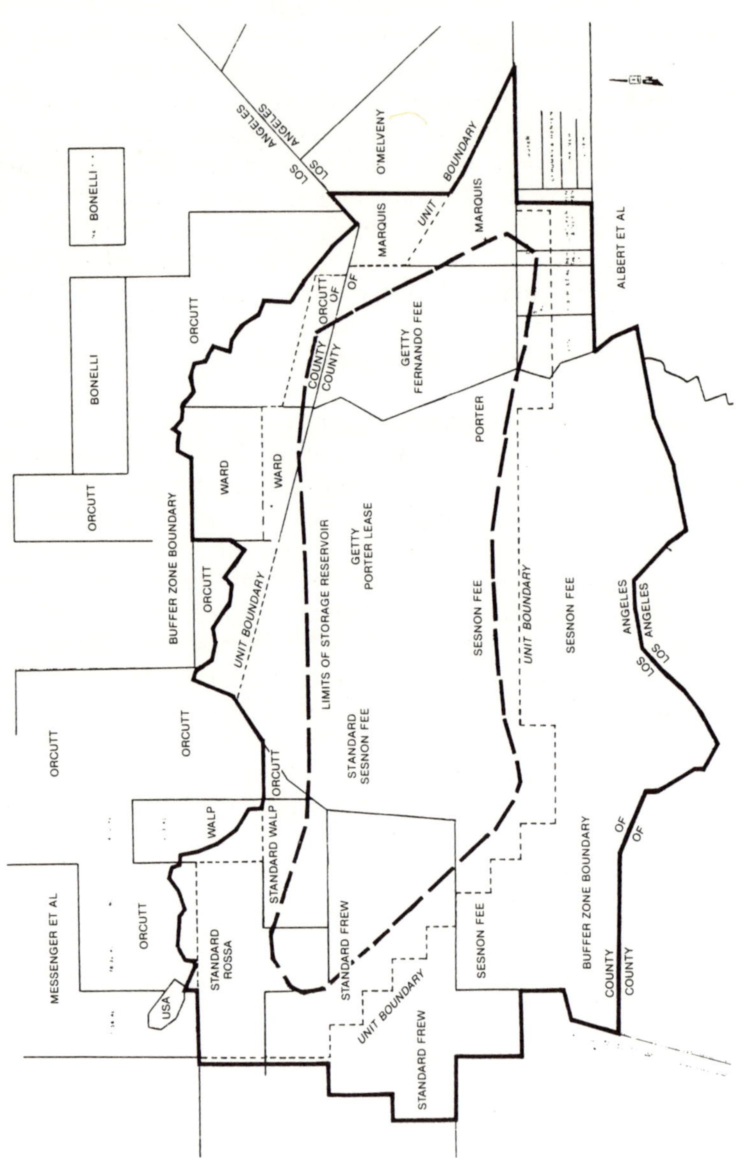

Figure 51. **Land boundaries of Aliso Canyon field.**

In conjunction with the 1972 Southern California Gas Company (SCGC) acquisition of the storage zone, approvals were required from several regulatory agencies. The California Public Utilities Commission approved the construction and operation of the storage project on February 23, 1972. Approvals for storage operations were also granted by the Los Angeles County Regional Planning Commission as well as the Los Angeles City Planning Commission.

The Department of the Interior declined to sell its interest in the federal leases. The storage field operator, therefore, entered into a contract with the department to pay a monthly gas storage fee. The department also retained its proportionate share of all oil produced from the Sesnon and Frew zones of the unitized area.

Royalty was also paid on the first 70 BCF (1.98×10^9 m³) of gas produced from the project. The governments royalty was 1.359 5 percent of the 70 Bcf deemed recoverable as of January 1, 1972.

RESERVOIR ENGINEERING

RESERVOIR SYSTEM

The selection of Aliso Canyon oil field as a gas storage field involved extensive reservoir engineering and geologic studies. A computer reservoir model was developed, and case studies were performed to assess the field's ability to meet projected seasonal and peak gas demands in the SCGC. An extensive amount of data was compiled to prepare these reservoir models, such as oil and gas production, historical reservoir pressures, well mechanicals, well logs, core analyses, and well tests. These data are summarized in Tables 7 and 7A.

For Aliso Canyon, the reservoir model worked well for verifying reservoir volumes and reservoir parameters in the oil belt. However, the estimated storage reservoir permeabilities in the gas cap were not as reliable because of the general lack of production of gas cap wells during primary oil depletion operations. Pressure drawdown and pressure buildup tests on selected wells, conducted before the field was acquired, aided in evaluating these gas cap permeabilities and in developing the reservoir models. Previous geologic studies indicated an unconformable surface and communication between the Sesnon and Frew zones. The concept of communication between the Lower Sesnon and Frew zones was simulated by assigning vertical permeability between the two zones across the northern area of the field. Formation oil volume factors and solution gas-oil ratios were calculated as a function of pressure from published correlations for reservoir oil and are shown in Fig. 52. Calculated gas formation volume factors and compressibility factors for the reservoir gas are shown in Fig. 53.

TABLE 7

Summary of Reservoir Characteristics of the Sesnon and Frew Zones of Aliso Canyon Oil Field
(English Units)

	Sesnon Zone				Frew Zone	
	Gas Cap		Oil Belt		Gas Cap	Oil Belt
	Upper	Lower	Upper	Lower		
Physical Properties of Reservoir Rock						
Porosity (bulk volume)	23.3	22.5	23.3	22.5	18.3	18.3
Permeability (mD)	118.0	74.0	118.0	74.0	62.0	62.0
Interstitial water						
(percent pore space)	24.0	25.0	27.5	30.0	28.0	35.0
Permeability variation	0.79	0.62	0.79	0.62		
Structural Features						
Subsea datum (ft)	5 150	5 150	6 000	6 000	5 150	5 860
Reservoir area (acres)	561	463	766	835	176	262
Gross zonal thickness (ft)	111	146	111	146	330	330
Average net sand thickness (ft)	40	62	58	57	65	87
Gas-oil contact, original (ft s.s.)	5 360	5 360			5 360	
Water-oil contact, original (ft s.s.)	–	–	Varies	Varies	–	6 360
Net sand volume (acre-ft)	22 241	28 557	44 442	47 366	11 468	22 756

(continued)

TABLE 7 (continued)
(Summary of Reservoir Characteristics of the Sesnon and Frew Zones of Aliso Canyon Oil Field
(English Units)

	Sesnon Zone				Frew Zone	
	Gas Cap		Oil Belt		Gas Cap	Oil Belt
	Upper	Lower	Upper	Lower		
Characteristics of Reservoir Fluids						
Gravity of oil (API)			22.4	22.4		21.0
Specific Gravity of gas	0.677	0.677	0.660	0.660	0.642	
Saturation pressure (psig)			3 258	3 258		3 230
Initial formation volume factor (bbl/bbl)			1.223	1.223		1.218
Initial solution GOR (cf/bbl)			462	462		447
Compressibility factor (z)	0.89	0.89			0.89	
Pressure and Temperature at Datum						
Original reservoir pressure (psia)	3 595	3 595	3 833	3 833	3 595	3 785
Reservoir pressure (1/1/71) (psia)	1 900	1 900	1 805	1 805	1 928	1 450
Reservoir temperature (°F)	174	174	188	188	174	185

TABLE 7A
Summary of Reservoir Characteristics of the Sesnon and Frew Zones of Aliso Canyon Oil Field
(Metric Units)

| | Sesnon Zone | | | | Frew Zone | |
| | Gas Cap | | Oil Belt | | Gas Cap | Oil Belt |
	Upper	Lower	Upper	Lower		
Physical Properties of Reservoir Rock						
Porosity (bulk volume)	23.3	22.5	23.3	22.5	18.3	18.3
Permeability (mD)	118.0	74.0	118.0	74.0	62.0	62.0
Interstitial water						
(percent pore space)	24.0	25.0	27.5	30.0	28.0	35.0
Permeability variation	0.79	0.62	0.79	0.62		
Structural Features						
Subsea datum (m)	1 570	1 570	1 829	1 829	1 570	1 786
Reservoir area (ha)	227	187	310	338	71	106
Gross zonal thickness (m)	34	44	34	44	101	101
Average net sand thickness (m)	12	19	18	17	20	27
Gas-oil contact, original (m s.s.)	1 634	1 634			1 634	
Water-oil contact, original (m s.s.)	–	–	Varies	Varies	–	1 939
Net sand volume (10^3ha-m)	2 743	3 522	5 482	5 843	1 415	2 807

(continued)

TABLE 7A (continued)

Summary of Reservoir Characteristics of the Sesnon and Frew Zones of Aliso Canyon Oil Field
(Metric Units)

| | Sesnon Zone | | | | Frew Zone | |
| | Gas Cap | | Oil Belt | | Gas Cap | Oil Belt |
	Upper	Lower	Upper	Lower		
Characteristics of Reservoir Fluids						
Gravity of oil (API)			22.4	22.4		21.0
Specific Gravity of gas	0.677	0.677	0.660	0.660	0.642	
Saturation pressure (kPa)			22 463	22 463		22 270
Initial formation volume factor						
(Rm³/SM³)			1.223	1.223		1.218
Initial solution GOR (Sm³/STB)			82	82		30
Compressibility factor (z)	0.89	0.89			0.89	
Pressure and Temperature at Datum						
Original reservoir pressure (kPA)	24 787	24 787	26 428	26 428	24 787	26 097
Reservoir pressure (1/1/71) (kPa)	13 100	13 100	12 445	12 445	13 293	9 997
Reservoir temperature (°C)	79	79	87	87	79	85

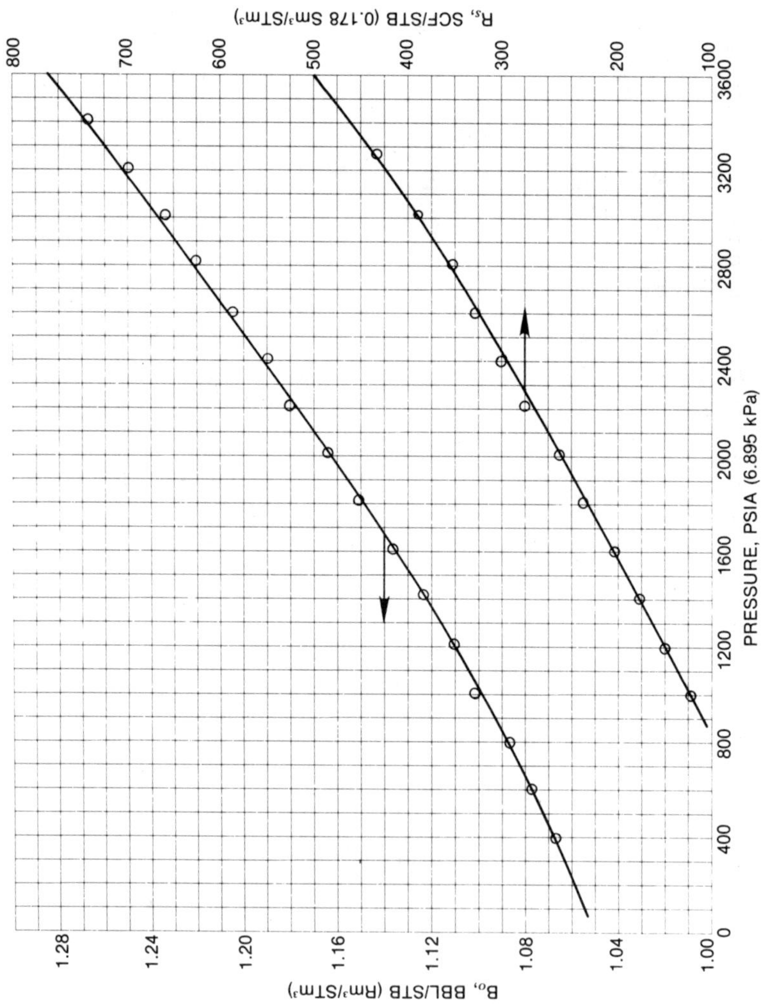

Figure 52. Plot of oil formation volume factor B_o and R_s ratio versus pressure, Aliso Canyon.

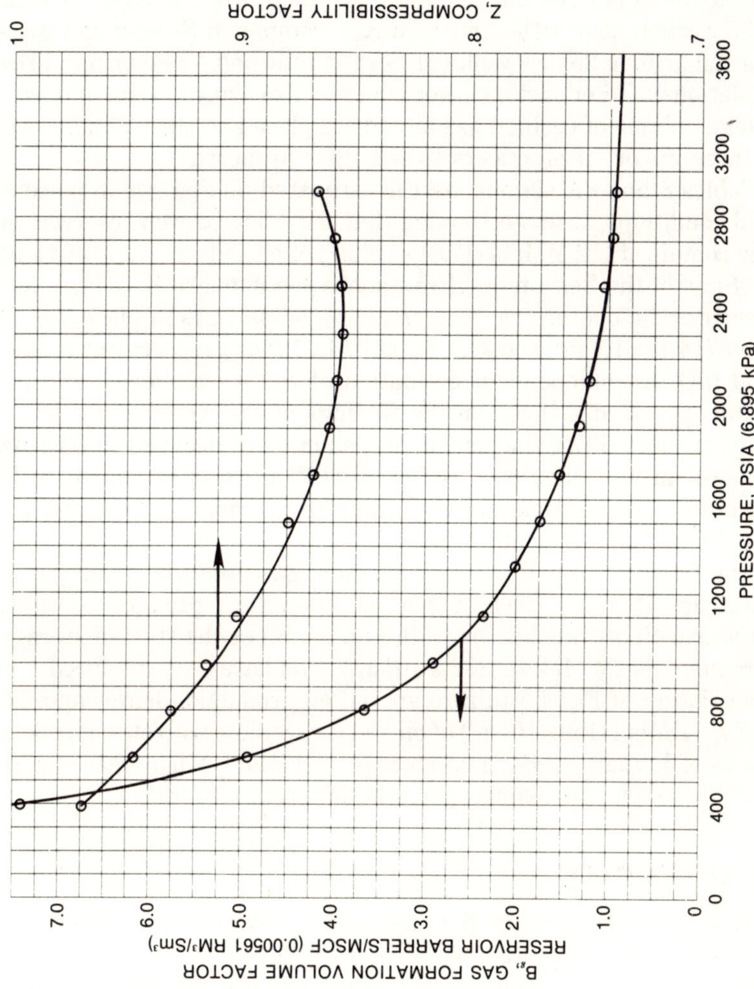

Figure 53. Plot of gas formation volume and compressibility factors versus pressure, Aliso Canyon.

Reservoir computer simulations indicated that Aliso Canyon was the most viable field for meeting peak and seasonal needs in comparison with other candidates. The reservoir model was a good tool for determining well and facility requirements and for predicting deliverability and injectivity at a range of inventories. It was predicted that deliverabilty of 1 000 MMcf/d (28.32×10^6 m³/d) could be achieved by reworking 35 Sesnon zone wells and 10 dually completed Sesnon and Frew wells and drilling 25 new wells (21 Sesnon zone and 4 Sesnon and Frew completions). Other factors favoring the Aliso Canyon selection were the high volume of cushion gas already in place and the willingness of the three major oil producers to sell their property.

Tables 8 and 8A summarize the estimated oil and gas in place at initial conditions. It was estimated that the Aliso Canyon reservoir initially contained 112 MMbbl (17.81×10^6 m³) of oil and 152 Bcf (4.304×10^9 m³) of gas in the Sesnon and Frew zones. On January 1, 1971, 1 year before storage operations, the remaining oil and gas in place was 91 MMbbl (14.47×10^6 m³) and 94.5 Bcf (2.676×10^9 m³), respectively (Tables 9 and 9A). An independent evaluation of the remaining oil and gas reserves provided a check on the findings of the reservoir model.

By January 1972, the remaining gas in place declined to 82.8 Bcf (2.34×10^9 m³); of this, 69.3 Bcf (1.96×10^9 m³) was estimated to be recoverable. At that time, the remaining recoverable oil was estimated to be 2.6 MMbbl (413×10^3 m³).

Fig. 54 shows a plot of oil production, gas production and gas oil ratio from 1940 to 1984. A plot of annual oil production versus cumulative oil provided is shown in Fig. 55. Since the start of storage, approximately 2.4 MMbbl (382×10^3 m³) of oil have been produced. The extrapolation of Fig. 55 to zero annual oil production leaves approximately 1 MMbbl (159×10^3 m³) of oil still recoverable with storage operations as of January 1985. Thus, storage operations will increase the recoverable oil by approximately 0.8 MMbbl (127×10^3 m³).

GAS INVENTORY

When the Aliso Canyon field was acquired for storage, the remaining gas in place was estimated to be 81.3 Bcf (2.30×10^9 m³). At that time the average reservoir pressure was 1 680 psia (11 583 kPa). From the reservoir studies performed earlier, it was determined that the field could be used effectively between reservoir pressures of 1 500 psia (10 342 kPa) and 3 400 psia (23 442 kPa). At 1 500 psia the remaining gas in place was estimated to be 72.5 Bcf (2.05×10^9 m³), and this estimate established the level of cushion gas for the field. The working capacity of the reservoir, or that volume of gas that could be injected

TABLE 8
Estimated Initial Oil and Gas in Place in Sesnon and Frew Zones of Aliso Canyon Field
(English Units)

	Sesnon Zone			Frew Zone	Sesnon and Frew
	Upper	Lower	Total		
Gas Cap					
Reservoir pressure (psia)	3 595	3 595		3 595	
Gas formation vol. factor (cf/scf)	444×10^{-5}	444×10^{-5}		444×10^{-5}	
Net sand volume (acre-ft)	22 241	28 557	50 798	11 468	62 266
Hydrocarbon space (bbl/acre-ft)	1 374	1 309		1 022	
Gas in place (Mcf/acre-ft)	1 737	1 655		1 292	
Gas in place (MMcf)	38 633	47 262	85 895	14 817	100 712
Oil Belt					
Reservoir pressure (psia)	3 833	3 833		3 785	
Net sand volume (acre-ft)	44 443	47 366	91 809	22 756	114 565
Oil formation vol. factor (bbl/bbl)	1.223	1.223		1.218	
SGOR (cf/bbl)	462	462		447	
Tank oil originally in place (bbl/acre-ft)	1 072	999		758	
Tank oil originally in place (Mbbl)	47,643	47 319	94 962	17 249	112 211
Solution gas in place (MMcf)	22 011	21 861	43 872	7 710	51 582
Gas Cap and Oil Belt					
Free gas in place (MMcf)	38 633	47 262	85 895	14 817	100 712
Solution gas in place (MMcf)	22 011	21 861	43 872	7 710	51 582
Total gas in place (MMcf)	60 644	69 123	129 767	22 527	152 294

TABLE 8A
Estimated Initial Oil and Gas in Place in Sesnon and Frew Zones of Aliso Canyon Field
(Metric Units)

	Sesnon Zone			Frew Zone	Sesnon and Frew
	Upper	Lower	Total		
Gas Cap					
Reservoir pressure (kPa)	24 787	24 787		24 787	
Gas formation vol. factor (Rm^3/Sm^3)	444×10^{-5}	444×10^{-5}		444×10^{-5}	
Net sand volume (ha-m)	2 743	3 522	6 256	1 415	7 680
Hydrocarbon space (m^3/ha-m)	1 771	1 687		1 317	
Gas in place (10^3 m^3/ha-m)	399	380		297	
Gas in place (10^6 m^3)	1093.87	1338.31	2 432.28	419.57	2 851.85
Oil Belt					
Reservoir pressure (kPa)	26 428	26 428		26 097	
Net sand volume (10^3 ha-m)	5 482	5 843	11 325	2 807	14 131
Oil formation vol. factor (Rm^3/STm^3)	1.223	1.223		1.218	
SGOR (Sm^3/STm^3)	82	82		80	
Tank oil originally in place (STm^3/ha-m)	1 382	1 288		977	
Tank oil originally in place (10^3 m^3)	7 575	7 523	15 098	2 742	17 840
Solution gas in place (10^6 m^3)	623.28	619.03	1 242.32	218.32	1 460.64
Gas Cap and Oil Belt					
Free gas in place (10^6 m^3)	1 093.96	1 338.31	2 432.28	419.57	2 851.84
Solution gas in place (10^6 m^3)	623.28	619.03	1 242.32	218.32	1 460.54
Total gas in place (10^6 m^3)	1 717.24	1 957.34	3 674.59	637.89	4 312.47

TABLE 9
Estimated Gas in Place in Sesnon and Frew Zones of the Aliso Canyon Field as of January 1, 1971
(English Units)

	Sesnon Zone			Frew Zone	Sesnon and Frew
	Upper	Lower	Total		
Gas Cap					
Reservoir pressure (psia)	1 900	1 900		1 928	
Gas formation vol. factor (Rcf/scf)	823×10^{-5}	823×10^{-5}		810×10^{-5}	
Net sand volume (acre-ft)	22 241	28 557		11 468	
Hydrocarbon space (bbl/acre-ft)	1 374	1 309		1 022	
Gas in place (Mcf/acre-ft)	937	892		708	
Gas in place (MMcf)	20 840	25 473	46 313	8 122	54 435
Oil Belt					
Reservoir pressure (psia)	1 805	1 805		1 450	
Net sand volume (acre-ft)	44 443	47 366		22 756	
Oil formation vol. factor (bbl/bbl)	1.149	1.149		1.129	
Gas formation vol. factor (Rcf/Scf)	870×10^{-5}	870×10^{-5}		1.11×10^{-5}	
SGOR (cf/bbl)	256	256		195	
Tank oil originally in place (Mbbl)	47 643	47 319		17 249	

(continued)

TABLE 9 (continued)

Estimated Gas in Place in Sesnon and Frew Zones of the Aliso Canyon Field as of January 1, 1971
(English Units)

	Sesnon Zone			Frew Zone	Sesnon and Frew
	Upper	Lower	Total		
Tank oil produced (Mbbl)	10 024	10 000		1 576	
Tank oil remaining (Mbbl)	37 619	37 619		15 673	90 611
Total hydrocarbon space (Mbbl)	58 267	57 871		21 004	
Oil reservoir volume (Mbbl)	43 224	42 880		17 698	
Estimated water encroachment (Mbbl)	2 500	2 500		–	
Free gas space (Mbbl)	12 543	12 491		3 306	
Free gas in oil belt (MMcf)	8 092	8 058	16 150	1 671	17 821
Solution gas in place (MMcf)	9 630	9 554	19 184	3 057	22 241
Total gas in oil belt (MMcf)	17 722	17 612	35 334	4 728	40 062
Gas Cap and Oil Belt					
Free gas in place (MMcf)	28 932	33 531	62 463	9 793	72 256
Solution gas in place (MMcf)	9 630	9 554	19 184	3 057	22 241
Total gas in place (MMcf)	38 562	43 085	81 647	12 850	94 497

TABLE 9A
Estimated Gas in Place in Sesnon and Frew Zones of the Aliso Canyon Field as of January 1, 1971
(Metric Units)

	Sesnon Zone			Frew Zone	Sesnon and Frew
	Upper	Lower	Total		
Gas Cap					
Reservoir pressure (kPa)	13 100	13 100		13 293	
Gas formation vol. factor (Rm^3/Sm^3)	823×10^{-5}	823×10^{-5}		810×10^{-5}	
Net sand volume (ha-m)	2 743.4	3 522.5		1 414.6	
Hydrocarbon space (m^3/ha-m)	1 771	1 687		1 317	
Gas in place (10^{3}m^3/ha-m)	215.1	204.8		162.5	
Gas in place (10^{6}m^3)	590.12	721.31	1 311.44	230.00	1 541.43
Oil Belt					
Reservoir pressure (kPa)	12 445	12 445		9 997	
Net sand volume (ha-m)	5 482	5 842.6		2 806.9	
Oil formation vol. factor (Rm^3/STm^3)	1.149	1.149		1.129	
Gas formation vol. factor (Rm^3/SM^3)	87×10^{-5}	87×10^{-5}		$1\ 111 \times 10^{-5}$	
SGOR (m^3/STm)	46	46		35	
Tank oil originally in place (10^{3}m^3)	7 575	7 523		2 742	

(continued)

TABLE 9A (continued)

Estimated Gas in Place in Sesnon and Frew Zones of the Aliso Canyon Field as of January 1, 1971
(Metric Units)

	Sesnon Zone			Frew Zone	Sesnon and Frew
	Upper	Lower	Total		
Tank oil produced(10^3m^3)	1 594	1 590		251	
Tank oil remaining (10^3m^3)	5 981	5 933		2 492	14 406
Total hydrocarbon space (10^3m^3)	9 264	9 201		3 339	
Oil reservoir volume (10^3m^3)	6 872	6 817		2 814	
Estimated water encroachment (10^3m^3)	397	397		–	
Free gas space (10^3m^3)	1 944	1 986		526	
Free gas in oil belt (10^6m^3)	229.14	228.18	457.32	47.32	504.63
Solution gas in place (10^6m^3)	272.69	270.54	543.23	86.56	629.79
Total gas in oil belt (10^6m^3)	501.63	498.72	1 000.55	133.88	1 134.43
Gas Cap and Oil Belt					
Free gas in place (10^6m^3)	819.26	949.49	1 768.75	277.31	2 046.06
Solution gas in place (10^6m^3)	272.69	270.54	543.23	86.56	629.79
Total gas in place (10^6m^3)	1 091.95	1 220.03	2 312.00	363.87	2 675.85

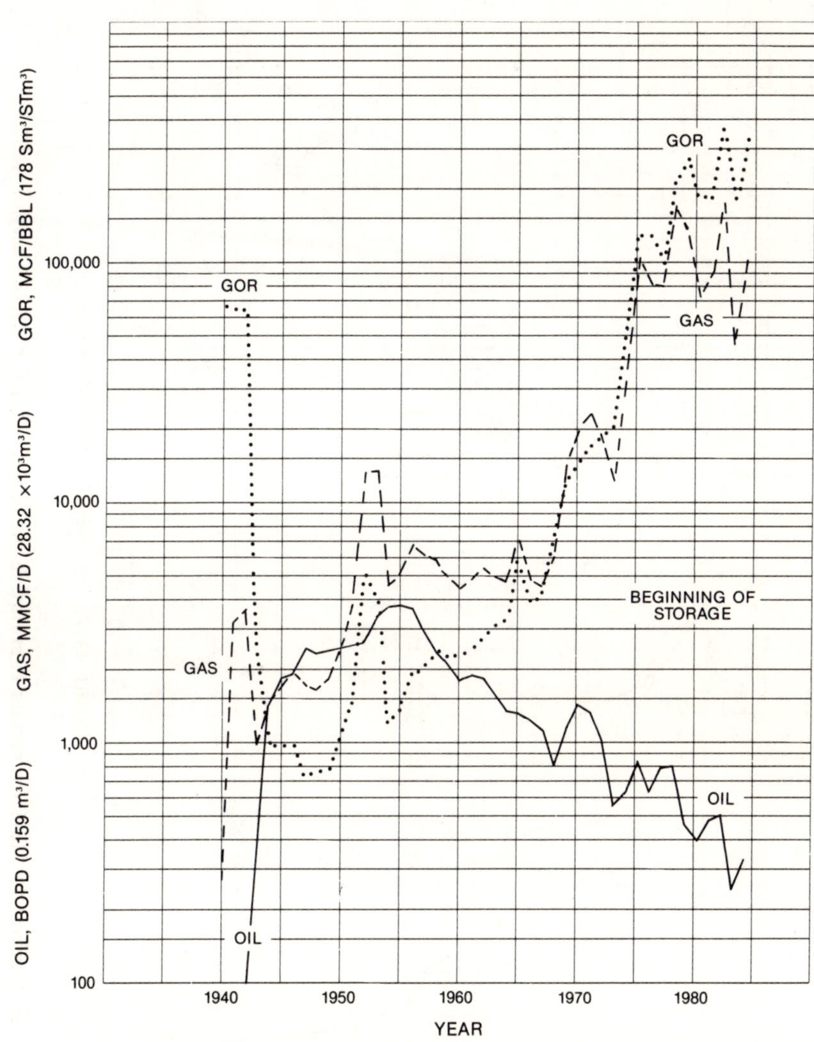

Figure 54. **Oil and gas production history of Aliso Canyon field.**

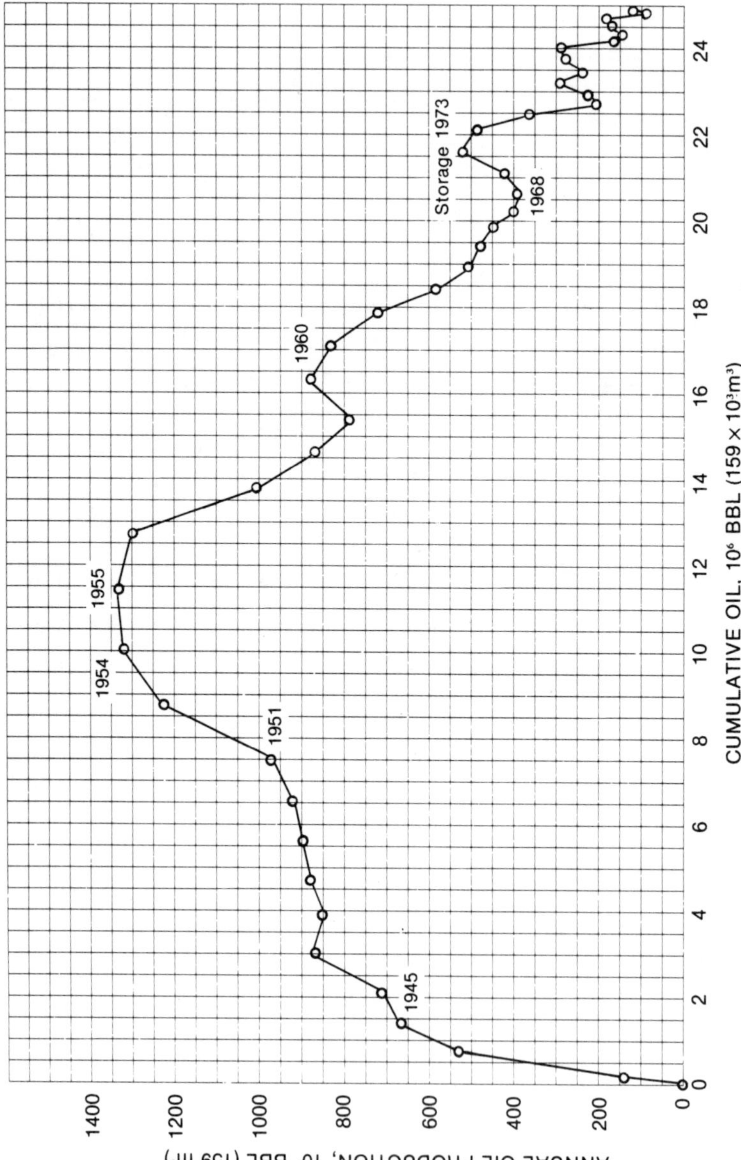

Figure 55. Annual oil production versus cumulative oil production for Aliso Canyon field.

or withdrawn between the operating pressure limits of the field, was determined to be 70 Bcf (1.98×10^9 m³).

As new information was learned about the reservoir system, the operating range increased from 1 175 psig (8 101 kPa) to 3 506 psig (24 173 kPa). Although raising the maximum reservoir pressure increased the total field capacity, the net result was an increase in the cushion gas volume; reaching the minimum pressure in all parts of the reservoir became unrealistic on account of the varying permeability across the rather large storage reservoir.

To track the gas inventory in the field, a hysteresis curve (plot of bottom-hole pressure versus gas inventory) has been maintained (Fig. 56). From 1973 to 1979 the hysteresis curve showed a shift to the right, indicating repressurization of the low permeability gas cap and some gas going into solution with crude oil. Because the reservoir has not been filled to its full capacity since 1983, the hysteresis curve has shifted to the left, indicating that gas is flowing out of tight regions and some additional gas is being liberated from the crude oil. Data to support the hysteresis curve are tabulated in Tables 10 and 10A.

DELIVERABILITY

Currently there are 99 withdrawal wells at Aliso Canyon with individual deliverabilities ranging from 1 MMcf/d to 79 MMcf/d (28×10^3 m³/d to 2.24×10^6 m³/d) at maximum working inventory. Back-pressure plots are maintained for each withdrawal well so that the productivity of the well can be closely monitored. These back-pressure tests are performed in conjunction with sand erosion tests to determine the maximum safe withdrawal rate for the well over the anticipated range of inventories. A sample back-pressure curve is shown in Fig. 57.

At Aliso Canyon, the aggregate well deliverability at maximum storage capacity exceeds 2 Bcf/d (56.63×10^6 m³); however, maximum field deliverability is limited to 1.5 Bcf/d (42.48×10^6 m³/d) because of facility restrictions. The maximum field deliverability is available from the top working inventory of 70 Bcf (1.98×10^9 m³) down to a volume of about 20 to 22 Bcf (566×10^6 m³ 623×10^6 m³). Deliverability below a volume of 12 Bcf (340×10^6 m³) is unknown because the full cycle volume of 70 Bcf has never been used to date.

SUBSURFACE FACILITIES

The discovery well for Aliso Canyon, Porter No. 1, was drilled in 1938 by the Tidewater Oil Company. All wells in the field were drilled

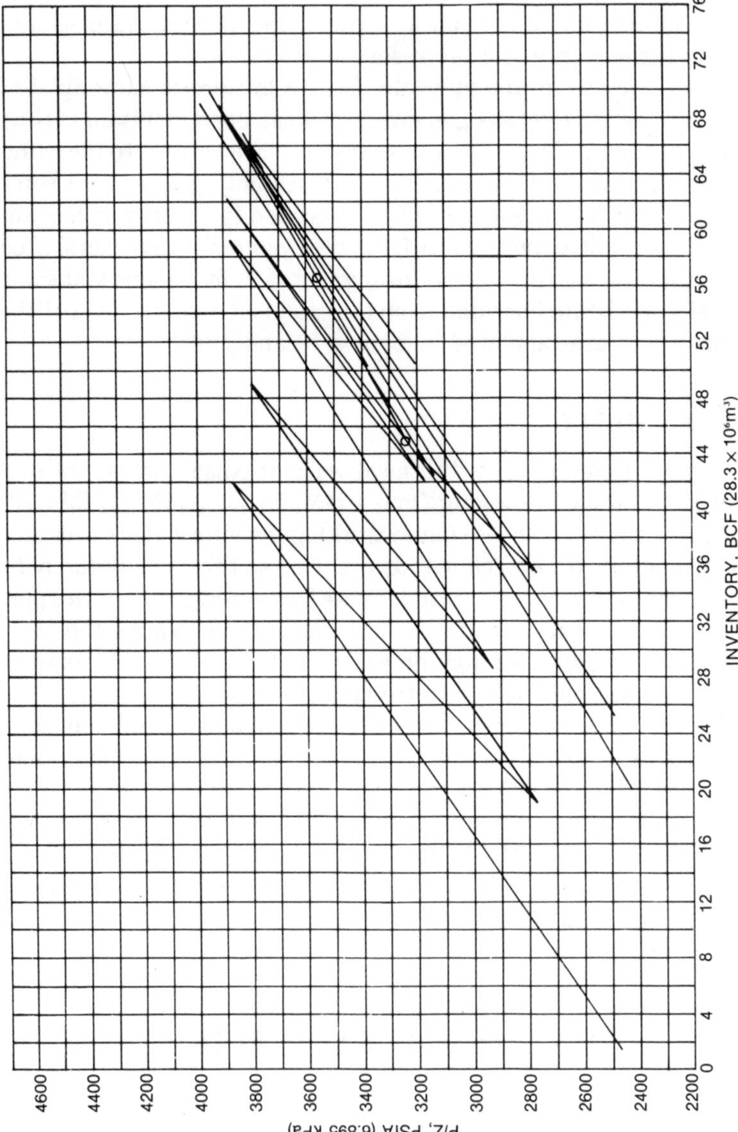

Figure 56. Plot of bottom-hole pressure versus gas inventory for Aliso Canyon storage field.

TABLE 10
Summary of Reservoir Pressures, P/Z, and Working Gas Inventory Since Start of Storage
(English Units)

Point	Date	Average Reservoir Pressures (psia)	P/Z (psia)	Inventory (Bcf)
1	1/10/74	2 149	2 471	1.5
2	10/19/74	3 421	3 889	42.1
3	4/22/75	2 400	2 773	19.0
4	9/19/75	3 351	3 819	49.1
5	1/22/76	2 539	2 937	28.7
6	11/24/76	3 430	3 897	59.4
7	2/07/77	2 751	3 180	42.0
8	9/15/77	3 444	3 911	62.5
9	3/12/78	2 674	3 093	40.7
10	11/30/78	3 539	4 002	69.1
11	4/13/79	2 112	2 427	19.9
12	11/28/79	3 505	3 970	70.0
13	3/31/80	2 912	3 392	50.4
14	11/03/80	3 373	3 841	67.0
15	4/01/81	2 786	3 219	50.4
16	12/02/81	3 457	3 924	68.9
17	4/19/82	2 167	2 493	25.3
18	9/24/82	3 506	3 971	70.0
19	3/17/83	2 402	2 775	35.3
20	11/28/83	2 800	3 235	44.5
21	11/10/84	3 107	3 569	56.8

TABLE 10A
Summary of Reservoir Pressures, P/Z, and Working Gas Inventory Since Start of Storage
(Metric Units)

Point	Date	Average Reservoir Pressures (kPa)	P/Z (kPa)	Inventory ($m^3 \times 10^9$)
1	1/10/74	14 817	17 073	0.042
2	10/19/74	23 587	26 814	1.192
3	4/22/75	16 547	19 119	0.538
4	9/19/75	23 104	26 331	1.390
5	1/22/76	17 506	20 250	0.813
6	11/24/76	23 649	26 869	1.682
7	2/07/77	18 967	21 925	1.189
8	9/15/77	23 746	26 965	1.770
9	3/12/78	18 437	21 325	1.152
10	11/30/78	24 401	27 593	1.957
11	4/13/79	14 562	16 734	0.564
12	11/28/79	24 166	27 372	1.982
13	3/31/80	20 078	23 387	1.427
14	11/03/80	23 256	26 483	1.897
15	4/01/81	19 209	22 194	1.427
16	12/02/81	23 835	27 055	1.951
17	4/19/82	14 941	17 189	0.716
18	9/24/82	24 173	27 379	1.982
19	3/17/83	16 561	19 133	1.000
20	11/28/83	19 305	22 305	1.260
21	11/10/84	21 422	24 607	1.608

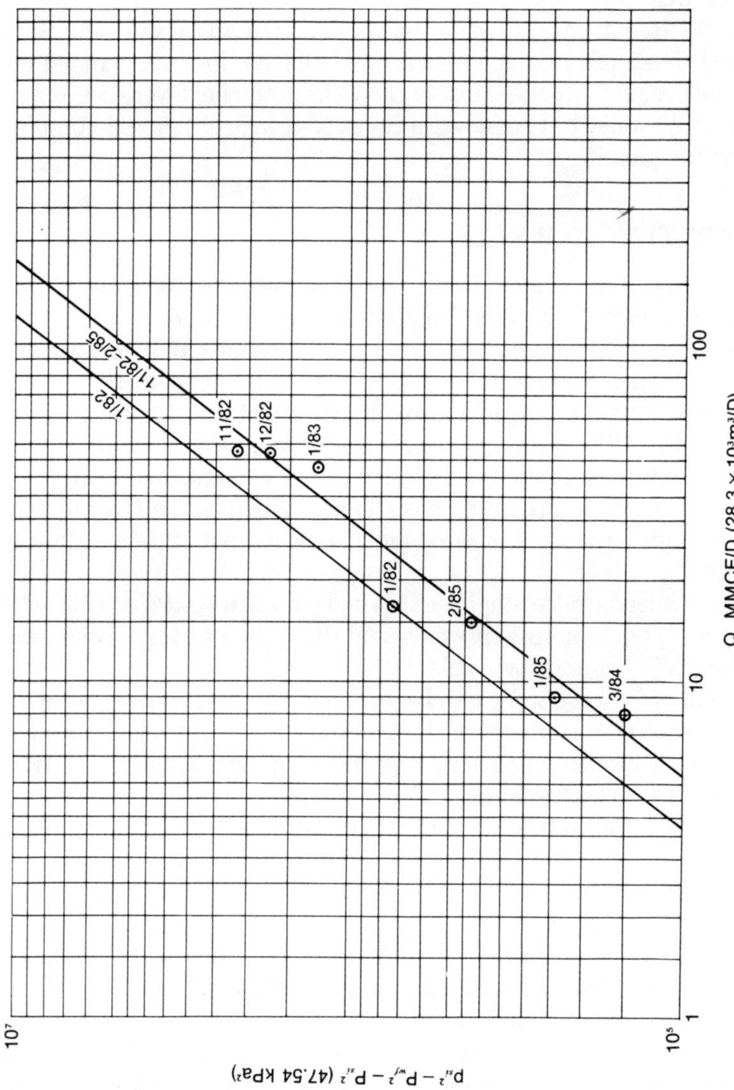

Figure 57. **Typical back-pressure curve for Aliso Canyon oil storage fields.**

with rotary equipment. Depending on the objective zone, drilling time varied from 20 to 190 days. Lost circulation was a problem in many wells, particularly during drilling through the large Santa Susana fault zone. Special drilling fluids were used to help prevent lost circulation; also, oil-emulsion muds were often used for drilling through the productive zones to reduce formation damage.

Typically, in a Sesnon or Frew zone well, 13–3/8-in. (33.97-cm) surface casing was used, mainly as anchorage for blowout prevention equipment (BOPE). The long or production string, either 8⅝-in. (21.90-cm) or 7-in. (17.76-cm) casing, was usually cemented above the zone to be produced. Most often, a shop-perforated slotted liner was landed opposite the production zone. Most of the wells in the field were completed as flowing oil wells, but some were completed with either rod-pumping units or for gas lift.

RECONDITIONED WELLS

Originally, Aliso Canyon wells were typically designed for oil production. A typical "old" well completion would include:

- An 800-ft (244-m) surface string of 54.5-lb/ft (81.10-kg/m), 13⅜ in. (33.97-cm), J-55-grade casing cemented to surface
- A production string of approximately 8 000 ft (2 438 m) made up of 7-in. (17.8-cm) long threads and couplings
- K-55 or N-80-grade casing with top of cement unknown
- A 5-in. (12.7-cm) production liner, cemented and perforated through the zone of interest and completed with 2⅞-in. (7.30-cm) tubing.

The wellheads were single-seal, 3 000-psi (20 684-kPa)-rating with small, ram-type, 2-in. (5.1-cm) valves on the tubing heads. These wells were typically tubing flow wells.

A total of 70 wells were purchased when the field was acquired, all of which required remedial work for gas storage use. A typical workover program for conversion to gas storage included the following steps (Fig. 58 illustrates a reconditioned well):

- Kill well with workover fluid, usually lease salt water. Move in and rig up.
- Install and test BOPE. (The California Division of Oil and Gas inspected and witnessed tests.)
- Pull and lay down production tubing and other equipment.
- Make up drill pipe and remove existing packer.
- Remove any junk left in well.
- Clean out and run casing inspection log, cement bond log, neutron log, and casing collar log. Pressure-test casing if necessary.

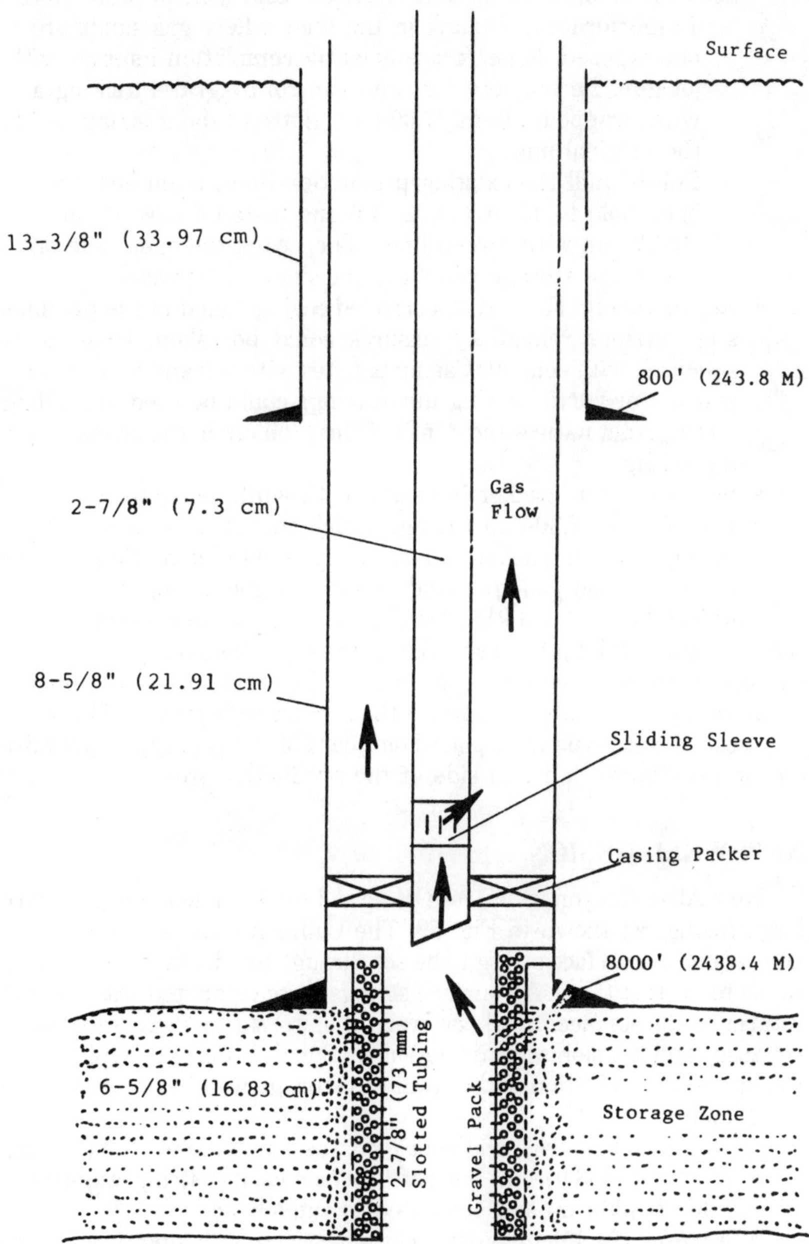

Figure 58. **Reconditioned Sesnon and Frew Zone storage well.**

- Provide sand-control protection. Two completion options were used for existing wells that required sand control protection:
  - Jet-perforate any blank in the liner where gas sands are not exposed. Reperforate existing completion interval with jet gun. Recomplete for sand control by gravel-packing a wire-wrapped, 2⅞-in. (7.30-cm), slotted tubing string inside the original liner.
  - Pull or mill the existing production liner, ream out the open hole to 13 in. (33.02 cm) and install a new 5½-in. (13.97-cm) wire screen liner. Then pack the open hole between the storage zone and the liner with gravel.
- Repair casing. For bad or corroded casing found in the previous step, various remedial measures could be taken. Holes were squeezed with cement. Casing patches with top and bottom seals and full or partial casing innerstrings could be used depending on the exact nature and depth of the problem in the primary casing string.
- Set production packer in casing with wireline equipment.
- Run 2⅞-in. (7.30-cm) tubing, which includes production tube through packer, packer seal assembly, packer latch, sliding sleeve run open, and gas-lift mandrel with pumpout plug above.

New 5 000-psi (34 473-kPa)-working pressure wellheads with casing seal flange and tubing head seal flange were installed on existing wells. Outlets on the wellhead are typically 3⅛ in. (7.94 cm). A pneumatic-actuated, fail-close valve is used for the surface safety valve. This valve is typically a gate-type valve placed outboard of the gate-type wing valve on the injection-withdrawal side of the production tree.

NEW WELL DESIGN

New Aliso Canyon wells were designed for high deliverability with large casing, as shown in Fig. 59. The California Division of Oil and Gas required a surface string to be set through any fresh-water-bearing sands to protect them. All surface strings were cemented with cement returns to the surface. In all new well tubulars, safety factors for burst, collapse, and tension were included in the well design. Typical well tubulars for an injection-withdrawal gas storage well were constructed as follows:

- A 20-in. (50.80-cm) conductor pipe was cemented to 40 ft (12.2 m).
- A 13⅜-in. (33.97-cm) buttress surface casing string was run to 1 000 ft (305 m) and cemented to the surface
- A 9⅝-in. (24.45-cm) buttress casing string with long threads and couplings was used for the production string and cemented from the top of the Sesnon zone to 3 000 ft (914 m).

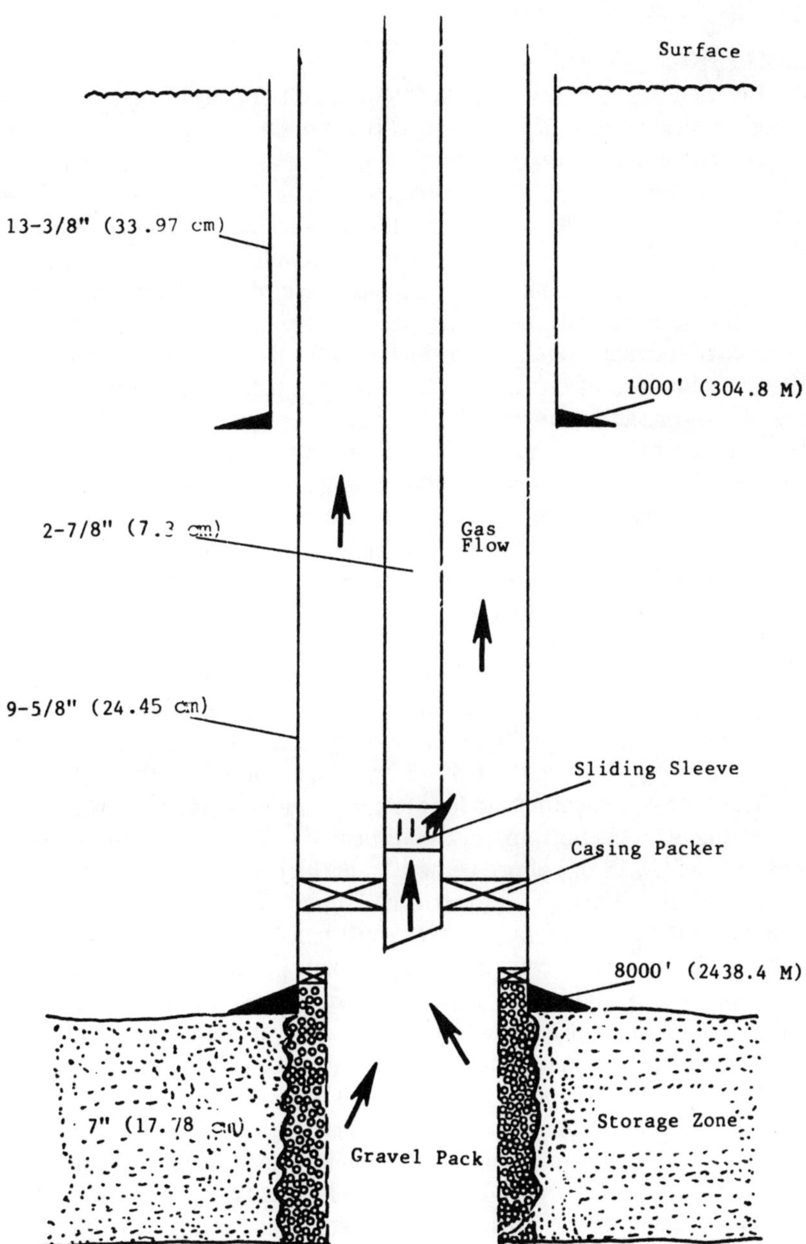

Figure 59. New storage well.

- A 7-in. (17.78-cm) wire-wrapped liner was set through the producing zone and completed, using open-hole gravel pack techniques.

DRILLING

Located in the Santa Susana Mountains, the Aliso Canyon Field is topographically one of the highest oil fields in California. The extremely rugged terrain has greatly increased drilling and production costs. It was originally decided that the new wells would be directionally drilled from multiple well sites with 15 ft (4.57 m) spacing between wellheads. Today, however, the preferred spacing between wellheads is 150 ft (45.7 m) whenever possible, to reduce the risk of a fire involving more than one storage well.

Two of the major design considerations for the drilling program were the probable loss of circulation to be encountered when crossing the Santa Susana fault zone and a cement shale program that could reduce the leakage of storage gas past the caprock along the outside of the wellbore. The gas company reduced drilling and mud costs substantially with the use of a nondispersed bentonite multiplex drilling mud. Physical properties of the mud were 18 c.p. plastic viscosity, a yield point of 10 lb/100 sq. ft. (0.49 kg/sqm), maximum gel strength of 2/10 lb/100 sq.ft. (0.098/.49 kg/sqm) and API fluid loss of 10 cc/30 min.

The bottom 700 ft (213 m) of casing was sandblasted and fitted with a float shoe on the bottom, a float collar on top of the third joint, and two centralizers and three scratcher clusters on the bottom four joints. Also one centralizer was fitted over every collar to 5 000 ft (1 524 m) and over every third collar from 5 000 ft to 1 000 feet (305 m).

The current program used for primary cementing is as follows. After the casing is on the bottom, it is reciprocated and circulated thoroughly. Then 500 cf (14.16 m³) of pre-cement mud flush are pumped, consisting of 0.25 lb/gal (0.429×10⁻³ kg/m³) of sodium acid pyrophosphate with 3 percent nonionic surfactant in fresh water followed with a calculated volume of 1:1 class G cement with pozzolan premixed with 2-percent gel, 1-percent dispersant, 0.2-percent HR-7, and 5-percent potassium chloride. This cement mix results in a lighter hydrostatic column, cleans mud cake from the formation, and provides better cement coverage between the casing and rock formations. The first mix is followed by a second mix containing 400 cf (11.33 m³) of class G cement mixed with 0.75- percent dispersant, 0.5-percent fluid loss, gas check, and 5-percent potassium chloride. Every effort is made to move the casing while displacing the cement without exceeding design yield factor.

Down-hole mud motors were used to establish borehole direction during drilling. All directional drilling was done in this manner. Wells were drilled to the top of the storage zone, where casing was set. Normally, a 12¼-in. (31.12-cm) hole was drilled. After the pipe was cemented, it was drilled out to within 10 ft (3.05 m) of the casing shoe and cement bond and neutron logs were run to confirm that cement bonding was satisfactory. Then drilling continued with 8⅜-in. (21.27-cm) bit to the base of the storage zone with clay-base drilling fluid. Resistivity, compensated neutron, and density logs were then normally run through the storage zone. Drilling mud was then changed over to HEC polymer for completion work. (HEC polymer is a low-solids completion fluid used to reduce fluid loss and contamination of the storage zone.)

The hole was opened up through the storage zone with reamers. Typically, the hole was opened up from 8⅜ in. (21.3 cm) to 15 in. (38.10 cm) for gravel packing. The gravel size most commonly used now is a 20- to 40-mesh.

CORING AND LOGGING

Several 3-in. (7.62-m) cores have been taken from the Aliso Canyon Field, most with a 7⅝-in. (19.37-cm) diamond core head with a plastic inner sleeve in the core barrel. Percussion-type sidewall core samples have been recovered in the gas cap for determining grain density, permeability, porosity, oil and water saturations, and oil-water ratios. Samples were selected for petrographic examination from cores penetrating the Sesnon zone. Depositional environment of strata, petrographic descriptions, modal compositions, textural characteristics, and lithologic description were determined for the main storage zone from this study for use in gas reservoir development. Core analysis was also used to determine clay content and type. X-ray diffraction analysis and cation exchange capacity data indicated clean sands (clay free). The conclusion was reached that Aliso Canyon sands are relatively clean (by California standards), finely grained feldspathic siltstone that contains mineral similar to clays but has not broken down to clay-size particles.

In addition, wet/dry sieve analysis data were collected on the storage zone. A core was obtained with a 3½-in. (8.89-cm) core barrel with rubber sleeve. The sieve analysis data were obtained by a combined wet/dry method, wherein clay-size particles were separated by disbursement, following the separation of larger particles, by means of sedimentation techniques. The larger particles were then separated by standard dry sieve analysis techniques.

Cores were cut and boxed at the wellsite and transported to the laboratory. Test plugs were cut approximately every 2 ft (0.61 m) for routine core analysis measurements. The residual core material was sealed with dipping plastic to preserve the fluid saturation for future tests. The test methods employed to obtain data followed the methods described in the API report on Recommended Practice for Core Analysis Procedure (API RP40) (see Appendix A). These methods consist essentially of

- Toluene extraction to remove oil and water
- The use of a Kobe porosimeter to measure porosity
- A flow tube-type permeameter for air permeability measurements

The logging program for the Aliso Canyon field included electrical, acoustical, and radiation logs on the entire hole. Cement bond logs were also run on the production casing to verify cement tops and integrity.

COMPLETION

Three basic types of completions have been used at Aliso Canyon. These are the open-hole gravel pack (Fig. 59), open hole with slotted liner or screen, and perforated completion with remedial sand control (Fig. 58). The open-hole gravel pack has proven to be the most successful method of subsurface sand control in new wells. In wells with low deliverability [less than 45 MMcf/d (1.274×10^6 m³/d)], gravel packs with slotted tubing have proven successful.

SAFETY DEVICES

Aliso Canyon incorporates several surface safety systems for use in case of emergency (see Fig. 60 for typical surface equipment and Fig. 61 for typical wellhead piping). These are

- Casing-tubing kill lines
- Fail-close pneumatic actuated valves on the withdrawal side of the production tree
- Low- and high-pressure pilots
- Fusible links
- Sacrificial sand probes

The casing-tubing kill lines provide a means of killing the well from a remote location 150 ft (45.72 m) from the wellhead in the event of either surface or subsurface blowout. These lateral pipelines are connected to the Christmas tree with standard gate-type wing valves installed. The kill line wing valves are left open at all times on the wellhead in case of emergency. The kill lines have block valves installed downstream to isolate wellhead pressure from the kill system.

The pneumatic-actuated valves, known as surface safety valves, provide the primary means of automatically shutting in a well when an abnormal situation is sensed at the wellhead. The surface safety valves

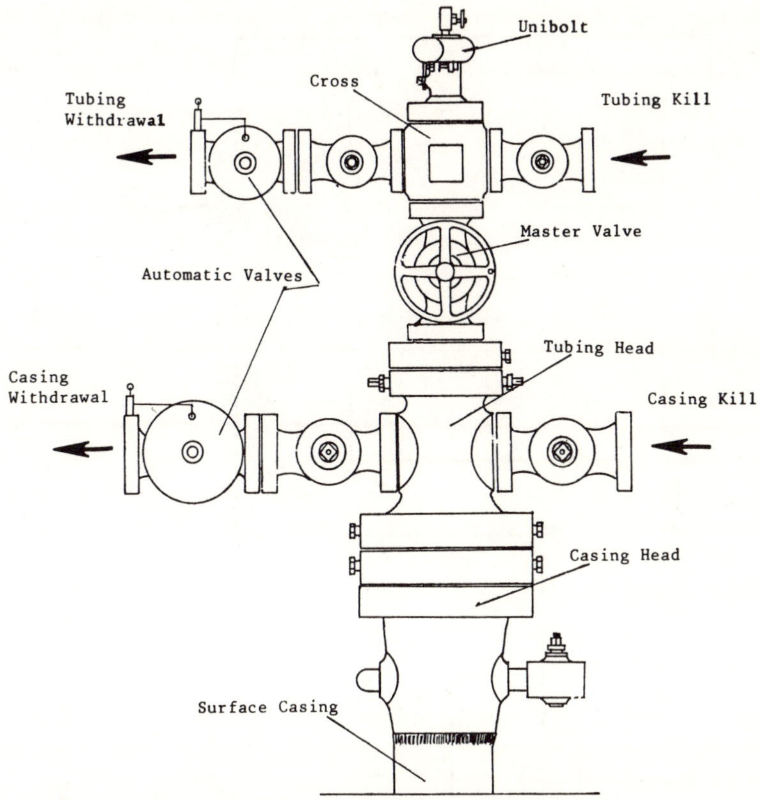

Figure 60. **Typical storage well surface equipment at Aliso Canyon storge field.**

can also be operated manually from a remote location. Once the surface safety valves are closed either manually or automatically, they remain closed until the problem is cleared and they are manually reset.

All storage wells have surface safety valves on the gas withdrawal side of the wellhead. Natural gas that is injected or withdrawn from the well passes through the surface safety valve.

During normal operations, the safety valves are kept open by an actuator. Pressure is applied to the actuator to keep the surface valve open. The actuator is operated by pressure of 100 psig (689 kPa), which is applied to or vented from the actuator through a pilot valve that is controlled by a 40-psig (276-kPa) control system. Before the 40-psig con-

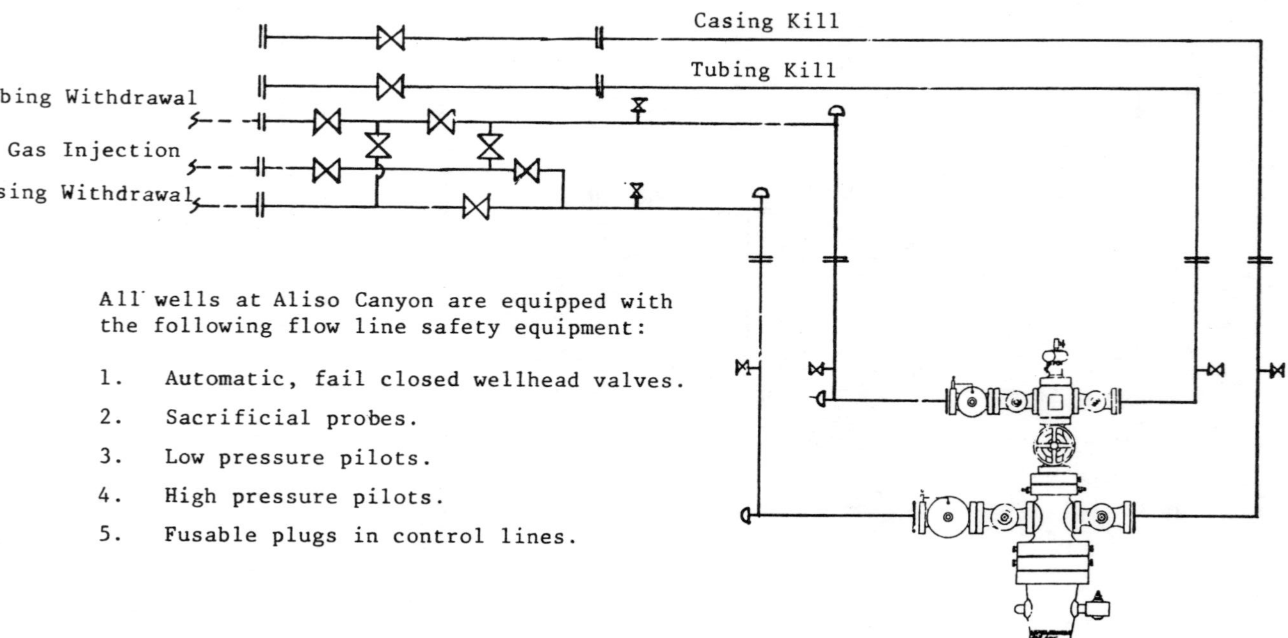

Figure 61. Typical wellhead piping at Aliso Canyon storage field.

trol pressure reaches the pilot valve, it passes through a series of sensors, any of which can cause the pilot to vent the 100 psig from the actuator to close the surface safety valve. These sensors are installed on the flow lateral pipelines of each well and continuously "sense" for fire, high and low pressure, and erosion of the sacrificial probe.

The sacrificial probe is a thin-walled sensor placed in the withdrawal flow line to prevent erosion damage to the wellhead and laterals. If the sacrificial probe's thin wall is eroded, the 40-psig (276-kPa) control pressure is vented, causing the pilot valve to vent the 100 psig (689 kPa) from the actuator, thus shutting in the well. The erosion of the sacrificial probe is generally caused by abnormally high reservoir sand or liquid production. All wells produce some liquids, and most wells produce sand in small quantities. However, when the reservoir rock or casing fails, the large amounts of sand produced quickly erode the sacrificial probe. The erosion of the probe shuts in the well before the flow line or wellhead is also severely eroded.

The fusible link is installed in the control pressure line. In the case of a fire at the wellhead, the fusible link melts and allows the control pressure to be vented. Once again, as the control pressure is vented, the pilot valve vents the actuator's 100 psig (689 kPa), closing the surface safety valve.

Flow-line pressure is constantly monitored by a high-pressure sensor, which is set at a predetermined high-pressure point. If this set point is reached, the 40-psig (276-kPa) control pressure is vented, causing the pilot valve to vent the 100 psig (689 kPa) from the actuator and closing in the well. The low-pressure sensor is identical in operation to the high-pressure sensor, but it is set for a low-pressure set point. A low-line pressure would be reached in the case of a ruptured flow line or sand-cut erosion.

WELL MAINTENANCE

Well maintenance involves the repair of casing leaks, water shut-off leaks or casing shoe leaks, and wellhead leaks. Casing leaks can be caused by the following:
- Old cement squeeze holes that have broken down
- Casing stage collar leaks
- Corrosion
- Drill pipe wear
- Tubular coupling thread leaks

These types of leaks may be repaired by squeeze cementing after obtaining a pressure breakdown with lease water and acid. Casing patches and partial inner casing strings are additional repair solutions.

Finally, large-diameter tubing flow strings can be run to minimize loss of deliverability. Use of innerstrings or large-diameter tubing isolates all high gas pressure from the suspect casing.

Gas leaks outside the casing shoe are thought to be caused by the pressure and temperature effects from cycling of the storage reservoir across various inventories. They can also be caused by poor primary cement, deteriorated cement, or leakage through original water shut-off holes in active or abandoned wells. A repair solution for active wells is to squeeze cement into the area of the casing shoe.

Wellhead seal leaks allow high-pressure gas to leak into the inner string, tubing, or surface casing annulus. Gas can then enter into shallow zones at the surface casing shoe or through casing holes. Possible solutions for this problem are as follows:

- Keep all annular pressures low enough to prevent gas migration into shallow zones by either connecting them to a low-pressure system or venting them to atmosphere.
- Install new wellheads with triple seals (use seal flange assembly between spools) on wells with obsolete wellheads when other well work is required.
- Inject plastic sealant to energize the seal in the head, and inject sealant to energize the seal in the seal flange.

SURFACE FACILITIES

The Aliso Canyon surface facilities consist of three compressor facilities, two dehydration plants, three liquid-gathering plants, a hydrocarbon recovery unit, field separators, field piping, an emergency shutdown system, and a methanol injection system (see Fig. 62 for a general flow diagram). Utility services for the station facilities are provided by gas engine-driven electrical power generators, steam-generating boilers, instrument and utility air compressors, starting air compressors, water systems, and a fuel gas system. The storage area covers approximately 4 000 acres (1 619 ha) of land with 101 gas storage and oil wells, 35 mi (56 km) of road, and many miles of pipe. A majority of the surface facilities were installed after 1972 except for one of the field compressor facilities, the liquid gathering plants, and the hydrocarbon recovery unit, which have been in operation since before World War II. These systems have been modified to handle gas recovered from the oil operation and process the gas for consumption.

In general, natural gas from various out-of-state supply sources is brought through SCGC transmission lines to the Aliso compressor station in late spring, summer, and early fall. Gas is received from the

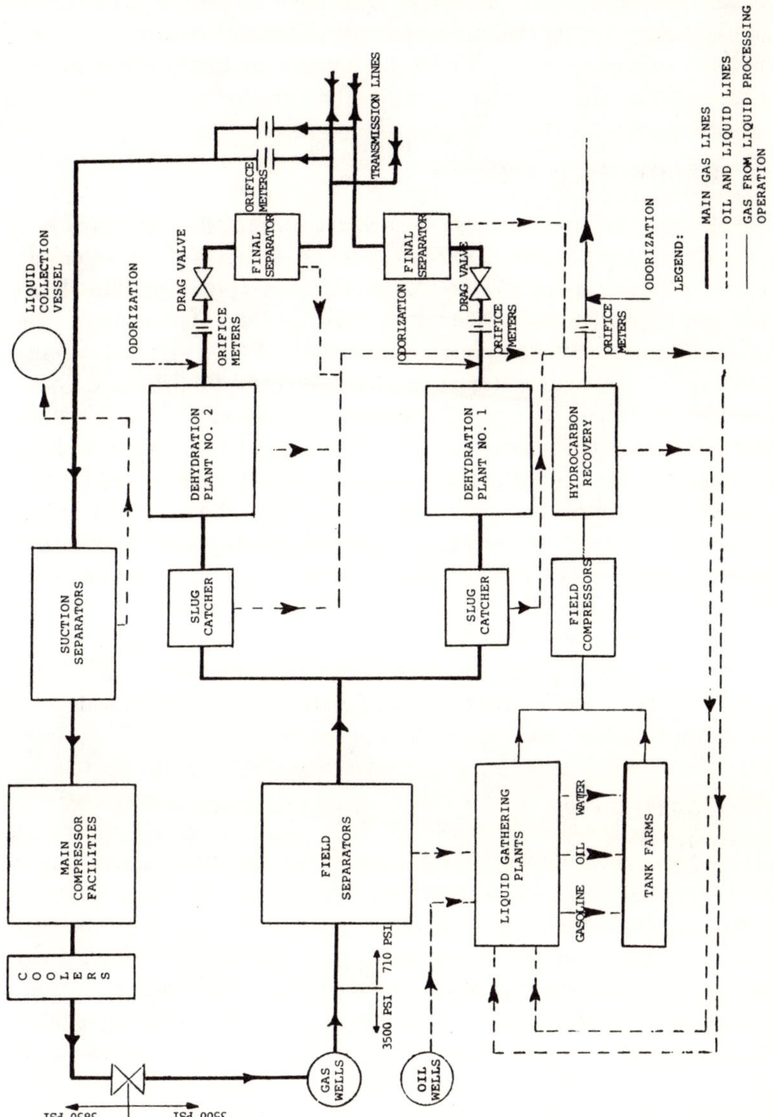

Figure 62. General flow diagram of surface facilities at Aliso Canyon field.

incoming transmission lines through two parallel injection metering tubes and gas compressor suction separators. After compression, gas is injected into the storage reservoir. Gas is withdrawn when demand is high – typically during the winter months, although heavy electricity generation loads can cause withdrawal operations during hot summer weather.

COMPRESSOR FACILITIES

The station has two compressor facilities totaling 42 700 hp (31 854 kW). The first facility, which is for gas injection, consists of six Ingersoll-Rand KVS 2 000-hp (1 492-kW), reciprocating, 300-rpm, two-stage, four-cylinder, gas-driven engines with air-cooled inter- and after-coolers. The second facility, also for gas injection, consists of three Dresser Clark 10 000-hp (7 460-kW), centrifugal gas turbine-driven compressors. These two facilities receive gas from the same suction source at a pressure varying from 250 to 600 psig (1 724 to 4 137 kPa) and compress it to a maximum of 3 500 psig (24 132 kPa). This pressure is set by the 3 500-psig design pressure of the injection piping. The combined injection capability, for the three gas turbine-driven and the six reciprocating compressors with 540-psig (3 723-kPa) suction pressure, can be as much as 500 MMcf/d (14.16×10⁶ m³/d). The third facility, which compresses gas recovered from the oil production operations for pumping into the company's intercity piping, consists of two Ingersoll-Rand XVG 2.5 MMcf/d (70.79×10³ m³/d) reciprocating field compressors located at Porter compressor plant.

FIELD SEPARATORS

The degree to which the injected gas will mix with residual crude oil and water present in the storage zone depends on the characteristics of the storage reservoir, velocity, pressure, and temperature conditions. Most of this liquid in the withdrawal gas is removed by 32 field separators. The field separators are vertical knockout drums or horizontal vessels, strategically located to remove the liquids from the gas flowing from each well or a well cluster before the gas flows to dehydration plants. Field separators are designed for 710 psig (4 895 kPa) at 150 °F (66 °C). They generally operate in the range of 500 to 645 psig (3 447–4 447 kPa) depending on the gas flow rates, amount of liquid entrainment, and distance from the dehydration plants. Each vessel is provided with vane-type elements designed for mist elimination. Gas enters and leaves at the upper part while liquid is removed from the bottom.

DEHYDRATION PLANTS

The station has two dehydration plants with a combined capacity of 1.5 Bcf/d (42.48×10^6 m³/d). The purpose of the dehydration plants is to reduce the water content of natural gas withdrawn from the wells to maximum water vapor of 15 lb/MMcf of gas (240.27 kg/10^6 m³), equivalent to a dew point of 47 °F (8 °C), and drop any hydrocarbon liquid that may have existed in the gas stream. The gas is dried by absorption of its water vapor in triethylene glycol. Water is removed from saturated glycol by boiling. After drying, the gas is measured, reduced in pressure, odorized, and sent to the transmission lines.

Gas Preparation

Gas withdrawn from the wells is prepared for dehydration by flowing through the field separators system to eliminate mist. If slugs of fluid move through the withdrawal system without being knocked out in the field separators, they are eliminated in the slug catcher vessel in the dehydration plant. The slug-free gas flowing out of the slug catcher vessel goes to the withdrawal gas cooler, where liquids condense.

Gas Drying

Gas and condensed liquids in the withdrawal cooler flow to two contactors operating in parallel. Each contactor has an entrainment separator in the bottom. Any liquid not separated out in the preceding operations is caught here and drained off. Gas, after flowing through the entrainment separator, flows upward through the chimney tray. The chimney tray also collects the water-rich glycol. The gas passes through seven bubble cap trays, where water vapor in the upflowing gas is absorbed in down-flowing glycol. Gas then passes through a demister tray and another entrainment separator before leaving the contactor.

Gas Metering and Pressure Regulation

Each dehydration plant has four meter tubes for measuring the withdrawal gas flow rate from each plant. After metering, the gas passes through a pressure-reducing station, i.e., drag valve. A separately instrumented bypass valve is provided to permit continued operation if a problem develops in the drag valve. Upstream pressure to the drag valve is set to maintain a maximum pressure on the contactors. The pressure downstream from the drag valve can vary from 250 to 520 psig (1 724 to 3 585 kPa) depending on the transmission system loop pressure. A low-signal selector switch positions the drag valve to satisfy

these pressure requirements. After pressure reduction, the gas flows
to final separators, where liquid that has not been removed by dehydration is knocked out from the gas stream. After odorization, gas flows
to the transmission lines.

LIQUID-GATHERING PLANTS

Two liquid-gathering plants, the hydrocarbon recovery unit (HRU),
the low-pressure field compressor, and the vent gas compressors normally operate continuously to process crude from oil wells and handle
liquids from dehydration field separators, plants, and drips (collection
devices) in low spots in the withdrawal piping system.

The two liquid-gathering plants, in the east and west areas of the
field, consist of several high- and low-pressure separators, surge tanks,
pumps, heater treaters, and oil and water storage tanks. For each gathering plant, the liquid from each oil well and from each field separator
is brought in to a common liquid production header. The production
header pressure is approximately 250 psig (1 724 kPa). From the header
the liquid flows to two high pressure separators operating in parallel
at about 200 psig (1 379 kPa). Gas liberated in the separators is metered
and flows into a high-pressure header before treatment in the HRU unit,
and the liquid flows into two low- pressure separators.

The low-pressure separators operate in parallel at about 65 psig (448
kPa). Gas that is flashed because of the pressure reduction in the low-
pressure separators is metered and flows to the low-pressure header,
where it is compressed and sent to the high-pressure header. From the
low-pressure separator, oil flows to the surge tank, where liquid surges
in the flow stream are eliminated. From the surge tank, oil is pumped
to the heater treaters operating at about 150 °F (65 °C) and 30 psig
(207 kPa). This heating breaks the oil-water emulsion and frees any gas
still in the liquid. Some water is separated in the heating section. In
the coalescing section of the heater treater the oil is separated from
the remaining water and is routed to oil product storage tanks. Any
gas recovered from this operation is directed into a low-pressure header.

HYDROCARBON RECOVERY UNIT

The HRU recovers the higher molecular weight hydrocarbons (i.e.,
gasoline) and water from the gas liberated in the liquid-gathering plants.
The unit capacity is 30 MMcf/d (849.5×10^3 m³/d) and consists of four
towers containing silica gel desiccant beds. Flow to each tower is
automatically controlled by either time or temperature cycle. At any
point in the cycle, two of the towers are drying the gas, a third is being

regenerated, and the fourth is cooling after regeneration. As the bed in a drying tower becomes saturated with hydrocarbon and water, the flow is automatically switched to a regenerated tower and the saturated tower is automatically regenerated. A fired heater provides regeneration heat.

The vapors driven off during regeneration are cooled below the dew point in an air-cooled heat exchanger. The liquid hydrocarbon and water are separated in a separator/surge vessel. The gasoline flows to a temperature-controlled stabilizing tower, where the desired vapor pressure is established. Vapors from the stabilizer are returned to the low-pressure header coming from the gathering plants. The gasoline is cooled in an air cooler, metered, and sent to a gathering plant, where it flows into the secondary flash drum and then to a gasoline storage tank. The dried gas from the tower is cooled, compressed, odorized, and sent to the transmission line.

EMERGENCY SHUTDOWN SYSTEM

There are two emergency shutdown (ESD) systems at Aliso Canyon: the main system and the fuel gas system. The main system, when triggered, isolates the station from inflow and outflow of gas and blows down the station piping through two blowdown headers and blowdown stacks by automatically closing and opening, in sequence, several ESD valves strategically located throughout the station.

The fuel gas ESD system, when activated, blows down the fuel gas for the gas-engine-driven electrical generators and other equipment. Triggering the main ESD system does not automatically activate the fuel gas ESD system; this system must be tripped manually. This arrangement allows the electrical generators to continue to operate and provide the emergency lighting necessary to assist station personnel in evacuating the compressor buildings and areas in the vicinity of the gas headers. Electrical circuits needed to protect equipment from damage also remain energized.

OPERATIONS

Staff groups in the Los Angeles headquarters of the SCGC plan the annual injection and withdrawal activities for all storage fields. Monthly withdrawal requirements are based on weather forecasts and demands from previous seasons.

During the withdrawal season, minimum working gas inventory targets are set for each month-end to meet both the seasonal and peak demand of the seasonal requirements. At Aliso Canyon storage field a minimum inventory of 20 to 22 Bcf (566×10^6 m³ to 623×10^6 m³) is held in the reservoir through the end of January to assure deliverability of 1.5 Bcf/d (42.48×10^6 m³/d) if an extreme peak day demand is experienced. Operating within staff guidelines, a gas control center is responsible for weekly and daily execution, including supply and demand considerations, weather forecasts, pipeline pack and draft, interruptible contracts, spot gas acquisition, and contract flexibilities.

Dispatchers from the gas control staff relay hour-to-hour withdrawal rate changes to the Aliso Canyon supervisors, who instruct the field operators to effect the rate change by manually opening or shutting in individual wells.

Gas is injected during the summer months, when the seasonal demands are generally low. Injection rates are usually more continuous than withdrawal rates, except when the weather is unusually warm and additional gas is needed to supplement the air conditioning load or to satisfy the demand of some of the industrial and commercial customers who must switch from oil to gas usage during heavy air pollution episodes.

INJECTION

Injection began in 1973 at the top of the structure on the east end of the field, where permeability was highest and the reservoir pressure was lowest. By injecting on the high structure first, it was hoped that the residual oil and condensate would be displaced down-dip.

After the annual storage cycles began, it was determined that pressure drawdown was more pronounced in the low-permeability areas of the field. To store all of the working gas during the injection season, it was necessary to begin injection in the low-permeability areas of the reservoir first and inject into the higher permeability wells as the reservoir pressure increased.

WITHDRAWAL

The first withdrawal season began in fall 1973. The initial plan was to flow the down-dip wells completed in the secondary gas cap first, then add the lower permeability wells completed in the primary gas cap, and finally, the high-permeability wells in the gas cap. On the first withdrawal test in 1973, the planned maximum rate of 1 Bcf/d (28.32×10^6 m³/d) was not reached. Problems included overload of the liquid-handling facilities, excessive sand production, subsurface safety valve failures, and hydrate blockage.

After 12 years of withdrawal experience, and modification of a number of surface facilities and down-hole devices, a majority of the initial difficulties associated with high withdrawal rates have been eliminated. A withdrawal rate of 1.5 Bcf/d (42.48×10^6 m³/d) can now be achieved if a minimum inventory of 20 to 22 Bcf (566×10^6 m³ to 623×10^6 m³) is maintained.

To control sand production and protect well component and surface facilities from sand erosion, maximum "sand free" flow rates from individual wells are determined by conducting frequent sand production tests on all gas storage wells. Doing so usually assures a flow of essentially sand-free gas from withdrawal wells.

RECYCLING AND OIL PRODUCTION

Recycling operations are performed to maintain or increase storage capacity by producing water, condensates, and crude oil from the lower part of the storage structure (see Fig. 63 for a schematic). Recycling operations have increased the gas-oil ratio from an initial average of 19 Mcf/bbl (3 384 m³/m³) during recycling operations in 1973, to 46 Mcf/bbl (8 192 m³/m³) in 1984. The number of wells used for recycling operations has been decreased as the gas-oil ratio has increased. At present the wells involved in recycling operations are limited to only 16 down-dip wells in the oil belt in the extreme southwest corner of the structure.

In addition to the revenues from oil production, recycling reduces the oil and condensate saturations in the gas cap and results in the reduction of liquid-handling problems during withdrawal operations. Recycling also increases gas saturation and relative permeability to gas, improving deliverability of the gas cap wells and reducing the tendency of the wells to load up with liquids.

INVENTORY VERIFICATION

Gas storage content is determined from the average reservoir pressure, temperature, and pore volume. Average reservoir pressures are determined from the shut-in tests conducted twice a year. For each test the field is shut in for at least 19 days, usually at maximum inventory at the end of the injection season and again at minimum inventory at the end of the withdrawal season.

Surface pressures of each gas well are measured with a deadweight tester once during each week of the shut-in period; the final set of readings is taken during the last two days of the shut-in. Pressures in down-dip wells that have a history of fluid levels are measured with an

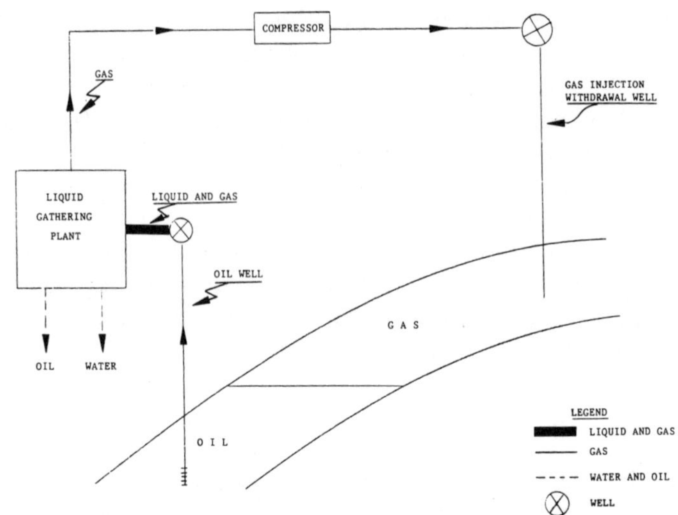

Figure 63. **Schematic diagram of recycling and oil production operations of Aliso Canyon field.**

Amerada-type wireline pressure gauge. Oil wells are not included in the computation because they do not necessarily reflect gas cap pressures.

TESTING FOR SAND AND DELIVERABILITY

The performance of the reservoir under various operating conditions is predicted by conducting well tests during injection and withdrawal periods.

The maximum sand-free flow rate for each well is determined from individual well sand tests. Before the test a sand probe is weighed and installed in the flow stream of the wellhead piping. After 6 hours of flow the probe is removed and reweighed. If weight loss is excessive, indicating that sand erosion may be a problem, the test is repeated with lower flow rates and smaller chokes until the sand probe weight loss is below the predetermined allowable limit. The flow rate established by the test is the maximum sand- free flow rate. Maximum sand-free rates are determined at various storage inventories so that the field deliverability can always be maximized by changing the chokes on individual withdrawal wells.

The maximum gas deliverability for the field under varying reservoir conditions is determined by correlating reservoir pressure against back-pressure production curves on the storage wells.

The maximum injectivity for individual wells is determined from tests at varying reservoir and surface pressures.

Gas, oil, and water flow rates are determined periodically for the oil wells by means of an automatic well test system.

Reservoir parameters in each well's drainage radius [e.g., conductance (kh), effective permeability, and skin] are calculated from pressure drawdown, pressure buildup, and pressure falloff tests.

WIRELINE PRESSURE AND TEMPERATURE SURVEYS

Temperature surveys are performed semiannually to determine reservoir integrity and identify any anomalous wellbore temperatures that may indicate well casing leakage or possible gas movement behind the casing. Fig. 64 shows a temperature survey plot for a typical storage well. The anomaly at 4 100 ft (1 250 m) reflects cooling due to a leak in the casing patch at that depth. Such leaks are usually verified with an acoustical noise log or radioactive tracer survey.

Pressure surveys on storage wells are performed routinely to determine shut-in reservoir pressure, flowing wellbore pressures, fluid levels in the wellbore, or gas and liquid gradients in the wellbore. Static bottom-hole pressure surveys are performed at the end of the high- and low-pressure shut-ins, usually twice a year. Fig. 65 shows a pressure survey plot for a typical storage well. The change in slope at 6 700 ft (2 042 m) indicates a fluid level within the storage zone in this well.

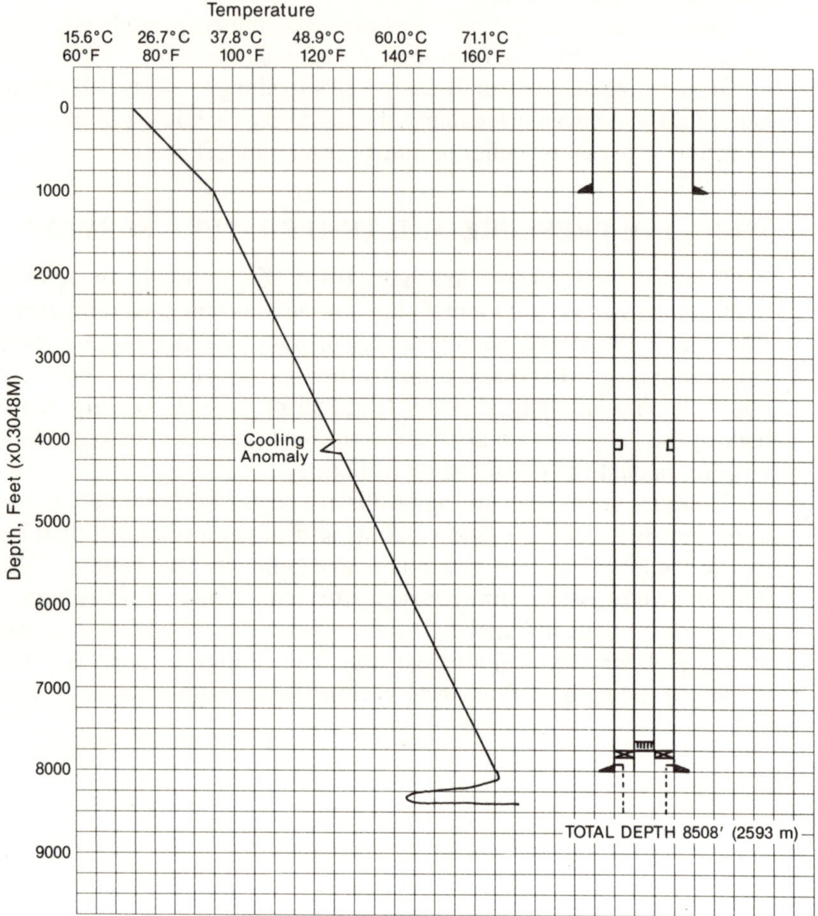

Figure 64. **Temperature survey plot for a typical storage well at Aliso Canyon field.**

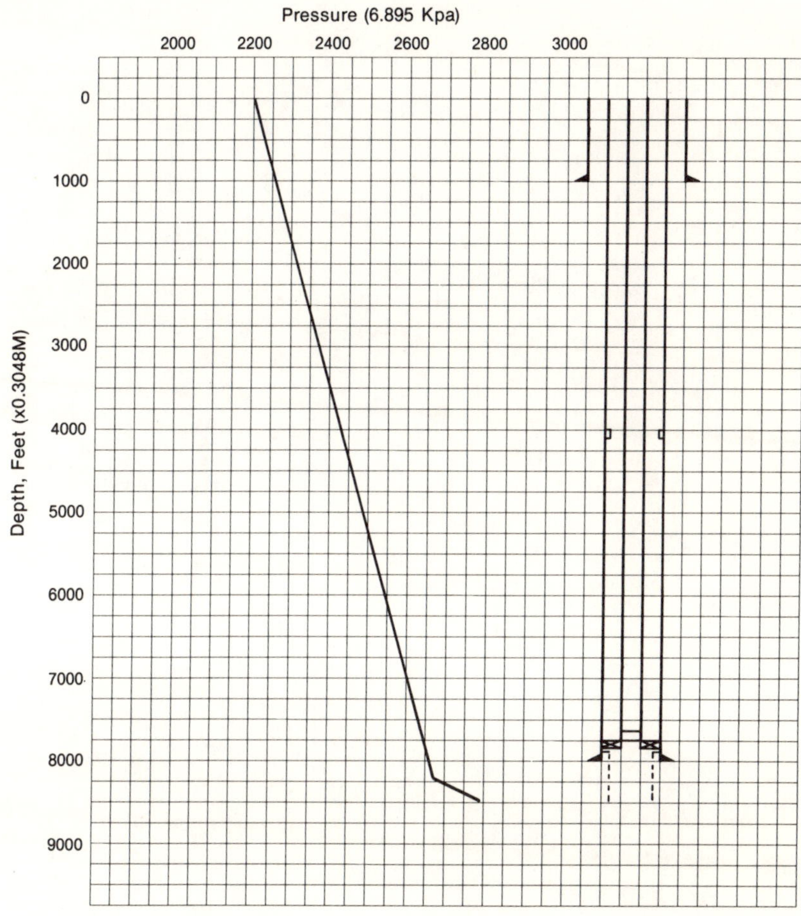

Figure 65. Pressure survey plot for a typical storage well at Aliso Canyon field.

APPENDIX A
CODES AND STANDARDS

GLOSSARY

BIBLIOGRAPHY

INDEX

Codes and Standards

AMERICAN PETROLEUM INSTITUTE (API)

1220 L Street, N.W.
Washington, DC 20005
Phone: (202) 682-8000

API 5A – Specification for Casing, Tubing and Drill Pipe

API 6A – Specification for Valves and Wellhead Equipment

API 5AC – Specification for Restricted Yield-Strength Casing, Tubing and Drill Pipe

API 5AX – Specification for High-Strength Casing, Tubing and Drill Pipe

API 5C1 – Recommended Practice for Care and Use of Casing and Tubing

API 5A2 – Bulletin on Threaded Compounds for Casing, Tubing and Line Pipe

API 1A – Specification for Oil Field Flat Belting (out of print)

API A10D – Specification for Casing Centralizers

API RP40 – Recommended Practice for Core Analysis Procedure

GLOSSARY

API – American Petroleum Institute

ANSI – American National Standards Institute (formerly American Standards Association)

annulus – the space between the outer wall of the pipe suspended in a well bore and the side of the open hole or the inner side of a larger pipe

aquifer – water bearing rock strata

barrel – a unit of measure to 42 gallons

baselogs – any number and type of logs which are recorded for reference as the original condition of the well and strata after completion

baskets – a type of tool used in the casing string for cementing purposes

bled-off – to release pressure from a well or a pressure vessle

blooey line – a pipe line connected to the drilling equipment and extended some distance from the rig through which the drill cuttings are blown out as dust in air and/or gas drilling operations

blowdown device – a mechanical control with which pressure can be systematically released

blowout – to go out of control as a result of a violent eruption of gas or gas and oil

blowout preventer – a device to control formation pressures in a well which operates by closing the annulus when pipe is suspended in the well and by closing the top of the casing at other times

blowout preventer controls – those devices which operate the opening and closing of the blowout preventer

blowout preventer rams – those parts of the blowout preventer which conform to the shape of the drilling tools and will effect a positive shut off when closed around same

bridging – to become plugged by debris lodged at some point in the well or cased bore hole

caprock – a comparatively impervious stratum immediately overlying the gas storage-bearing rock

casing – a term applied to steel or iron pipe used in a well to prevent caving of the walls or ingress of water, or both

casing, fiber stress – is the resultant internal force that resists change in the size or shape of a body acted on by other forces

casing, grade – an established standard whereby casing is rated according to its chemical and physical properties

casing, injection-withdrawal – refers to the last casing string in the well through which gas is injected and withdrawn

casing, minimum collapse strength – is the minimum collapse strength prescribed by the specification under which pipe is purchased from the manufacturer

casing, minimum yield strength – is the minimum yield strength prescribed by the specification under which pipe is purchased from the manufacturer

casing, running – the procedure of connecting each joint of casing on the rig floor and lowering same into the well bore

casing, tensile strength – is that tension strength prescribed by the specification under which pipe is purchased from the manufacturer

catheads – a winch head on the draw-works countershaft in a rotary rig

catline – a line smaller than the heavy cables used on well equipment that is used to raise and lower lighter tools and equipment about the rig during drilling operations

cement bond – the adhesion of the cement to the casing and wall of the hole

centralizers – a tool run on the casing string to hold the casing in the center of the hole

chemical goggles – virtually air tight goggles

christmas tree – an assembly of valves and fittings located at the casing head of an oil and/or gas well for the purpose of controlling the flow of gas or oil from the well

circulating time – that time element required to complete one circuit of the hole displacement

collar float – a collar used on casing during cementing operations. The backpressure valve with which it is fitted permits casing to be floated into the well and serves as a stop for the cementing plug

commingling – the ability of two or more compatible substances to join as a unit compressive strength – the unit compressive force that a material will stand prior to failure

cycle, closing – an operation wherein a device ends up in a closed position, such as a valve

cycle, opening – an operation wherein a device ends up in an open position, such as a valve

degasser – a device to free gas from liquids

delta pressure – the act of raising the pressure of a formation above the original or discovery pressure

density – overburden – the specific gravity or relative density of the overlying rocks or substances

derrick man – a worker in a drilling or producing crew who racks pipe in the derrick of a rig

dogs – tempered steel pieces which grip and hold certain tools, drill pipe, etc. in place

draw works – the collective name for the hoisting drum, shaft, clutches, and other operating machinery used in the drilling of a well. Draw works serves as a power control center for the hoisting gear and rotary table

drilling – an operation whereby a vertical hole is made from the surface down to a specific horizon level. Two systems generally used are percussion and rotary

drilling logs – a detailed drilling record which gives color, nature, thickness and content of the formations encountered

drill pipe – a pipe of the best quality seamless, alloy, heat treated steel used in an auger stem in drilling a rotary well

drill pipe, laying down – to pull drill pipe out of the hole in single lengths and lay it on a rack

drill stem test – a test to determine the size of the flow of a well by attaching a drill stem tester to the drill stem so that the medium from the producing sand can flow into the pipe

drum – a 55 gallon metal container, a winding cylinder with flanges on which a drill or hoisting line is wound

fee ownership – land operated for oil or gas under a right of ownership as contrasted with land operated under a lease

flow potential, well – the estimated yield or flow capacity of a well for a specific time under specific pressure conditions

gas, entrained – a substance carrying gas along with it

gas, sour – a gas having excessive sulphur content

gas, storage – the placing of gas in a geological formation by mechanical means for future use

gas, sweet – a gas which is comparatively free of sulphur

gas, toxic – a gas which is determined to have a detrimental effect on living organisms

geological sections – those layers of rocks which make up the earth's subsurface and are classified by time

geophysical programs – programs designed to use devices based on principles of physics for the purpose of determining subsurface structures

hole displacement – the volumetric content of the hole

hole, dry – usually referred to as a hole which is void of oil and/or gas

hydrocarbons – that series of alkane hydrocarbons generally found in petroleum, including methane, propane, butane, pentane, etc.

hydrogen sulfide – a toxic compound found in sour gas hydrostatic head – the weight of a liquid column computed by density and height in pounds per square inch

hydrostatic pressure – the pressure caused by the weight of fluid which fills the pore space of the host rock

I/W – injection/withdrawal

injection – the placing of a fluid into storage

inhibitor – a substance which slows or interferes with a chemical reaction (such as corrosion)

joint, swing – a connection which permits or allows a change in direction

kill – to stop the flow from a well by filling it with liquid

latent defects – an imperfection which is present but not visible to the eye

lease – a contract for the possession or use of lands for a determined period or consideration of payment or rent

lines, treating – lengths of connected pipe through which fluids are pumped for treating purposes

logging – a method of determining subsurface information by mechanical, electrical or sonic means

logs, caliper – a log used to determine the diameter of a hole

logs, correlation – logs which may be related one to another

logs, downhole casing inspection – logs used to determine the integrity of the casing logs, neutron – logs used to determine hydrogen content of the strata which is obtained by running a device in the hole

lubricator – a device temporarily mounted above a mastercontrol valve which houses tools to be run into and pulled back out of the hole under pressure

map, gathering system – a sketch of the pipelines and associated equipment connecting the wells to the compressor station or market lines

mining operations, subsurface – physical mining of material through tunnels or shafts under the surface by men and machinery

mining operatings, surface – physical mining of material by moving the surface by men and machinery

mud – drilling liquids circulated in the hole for removal of cuttings, cooling, lubrication, and providing protection from blowouts

mud cake – a filter cake formed on the walls of the hole to prevent the loss of liquids in the hole

mud gun – a nozzle through which liquids under pressure are passed to mix the drilling liquids circulated in the hole

mud pit – a container for the mud usually an open pit in the ground or a steel vessel

mud pit level indicators – devices which indicate the amount of mud contained in the pit or vessel

perforating tool – a device (a mechanical instrument normally containing an explosive) which is lowered into a well and discharged at a desired depth to open the casing so that the gas may be injected and withdrawn from the formation

perimeter of field – a line that establishes the estimated outside boundaries of a field including both the reservoir and buffer zone

permeability – a measure of conductivity of a rock to the movement of fluids through it, expressed in darcies or millidarcies

pilot light – a small continuous flame which ignites a larger intermittent flame

plugs, cement – a device made of various materials which is pumped down the casing behind the cement and seated against a holder which will effectively shut off flow back of cement

pore space – the space between the grains of sediment in a clastic rock or the space caused by chemical reaction in a nonclastic rock

pressure, abnormal – that unit of force exerted which is either substantially higher or lower than the normal hydrostatic gradient at relative depth

pressure, fracturing – the unit of force required to physically part the rock

pressure, low vapor – a minimal force caused by the constant motion of the molecules in a fluid and their impact on each other which tends to make the molecules fly apart

pressure, maximum reservoir – that unit of force which is established as the highest possible pressure

pressure, threshold – that pressure which initiates first movement of the connate water contained in the caprock

raceways – canals for liquids

reciprocating – to move forward and backward alternately

recondition – to re-enter a completed bore hole for the purpose of additional work, to remove impurities and/or add new substances to mud

records, scouting – records which are obtained by industry personnel exchange, usually pertaining to new wells being drilled

regulator–a device which controls pressure by reduction

relief lines–lines of pipe and valves which allow one to control the amount of pressure considered safe or desirable

remote controls–controls which are maintained a fixed distance from the area of the units they control

reservoir, analysis–the study or determination of the nature of the reservoir

reservoir, areal extent–the geographical extent and configuration of the porous rock which serves as a gas depository

reservoir, geometry–the physical shape of the porous rock which serves as a gas depository

reservoir, storage leakage–any gas which escapes from the porous rock depository by an avenue that is not controlled

reservoir, stratigraphy–a geological interpretation which treats the formation composition, sequence, and correlation

rig–all the drilling and pumping equipment, including the derrick, used in drilling a well

rig site–a prepared location upon which the rig and associated equipment are assembled

rig up–assembling of machinery and tools in position and condition for drilling

rotating–the turning of the turntable used to rotate the pipe in a rotary rig

safety equipment, downhole–devices installed in the producing casing which will automatically close and seal off the flow should a failure exist above the setting depth of the device

scratchers–a device attached to the casing used to scratch or scrub the walls of the well

sloughing–the act of the side walls of the bore hole moving into or falling into the well bore from the existing wall

spinning rope–a line powered off the cathead used on some rigs in making up drill pipe and casing

stand–a unit length of drill pipe or tubing stacked in the derrick

standpipe–the vertical pipe rising along the side of the derrick and joining the mud pump to the rotary hose

staging tool – a device, run in the casing string, used in cementing the casing in a well which due to its design allows the cement to be placed at different depth intervals at different times

storage, area – that area within the perimeter of the field which contains the gas

storage facilities – the physical entities involved in the storage of gas such as lines, compressors, wells, and the gas storage formation

storage, protective area – that area surrounding the outerlimits of the reservoir which is deemed necessary to protect the reservoir from encroachment and from other operations that may affect the integrity of the reservoir

storage, reservoir – the porous formation in which the gas is stored

storage, reservoir integrity – the ability of the storage formation to contain the gas in place

storage reservoir, interval – the thickness in feet of the gas storage rock

swabbing – the act of lifting fluids in the casing of a well to the surface by means of a steel and rubber device which operates on a cable

tear down – to disassemble the machinery and tools used in the drilling of a well

tight hole – a well for which no information is released

tongs, hydraulic power – hydraulically operated wrenches used to connect or disconnect threaded tubular goods

toolpusher – a foreman who oversees the rig operations

tour – a work shift; often pronounced "tower"

trip – either pulling a string of pipe or tools out of the hole or running a string into the hole

tubing rams – devices of a configuration which will allow them to close around the outside of the tubing in such a manner as to prevent any flow of gas or oil on the outside to pass between the tubing and casing

tubing string – a string of small diameter pipe, usually the last string run in a well

valve, check – a valve which permits flow in one direction but automatically closes when flow is attempted in the opposite direction

valve, master – principal well head control valve

valve, safety–a valve that automatically senses and controls the pressure in the well and which is set to ensure that the maximum safe pressure of the well is not exceeded

valve, wing–a valve mounted on the side of the well head above the master valve for the purpose of controlling the flow of gas or oil

well head–an assemblage of equipment attached to the casing at ground level that is used to control the flow of the well

well head control equipment–that part of the well head assemblage that is used to control the pressure and flow

well kicking–a well that surges intermittently

wells, collection–a well completed in an upper zone above the main storage zone to collect any gas that might escape from the storage zone

wells, observation–a well completed in the storage area used to monitor the integrity of the storage zone or other zones

wrap–a single coil of rope around an object such as a cathead

zone–a term used to distinguish sections of interlayered sand, shale, chert, and other rocks

zone, buffer–an area lying between the outer limits of the effective gas storage zone and the periphery of the protected area

zone, collection–a rock strata located above the storage zone which any escaping gas migrates

zone, creep (plastic flow)–a rock strata which moves

zone, lost circulation–a rock strata which, due to crevices or extremely vugular porosity, will absorb the drilling fluid rapidly

zone, unconsolidated–a rock strata which is made up of a physically loose, caving material

zone, upper gas–a rock strata above the main storage zone that contains gas

BIBLIOGRAPHY

American Gas Association, *Supplemental Gases—Peak Shaving/Base Load*, Gas Engineering and Operating Practice Series, Book S-2, Arlington, VA, American Gas Association,1987, 502 pp. [Catalog No. XY8701]

Barbour, Joel S. and David Cummings, "Applications of Downhole Borehole Video Cameras for Geotechnical and Geophysical Investigations of Underground Gas Storage Facilities," *A.G.A. Operating Section Proceedings 1987*, American Gas Association, Arlington, VA, 1987, p. 566.

Berkstickler, A.C. (ed.), *Symposium on Salt*, Northern Ohio Geological Society, Cleveland, 1963, 661 pp.

Boulanger, A. and J.P. Lagron, "LNG Storage in Deep-Seated Caverns," *A.G.A. Operating Section Proceedings 1984*, American Gas Association, Arlington, VA, 1984, p. 157.

Bowden, Joe R., "'Are Your Prepared?" *A.G.A. Operating Section Proceedings 1983*, American Gas Association, Arlington, VA, 1983, p. T-326

Brown, L.W., "Abandoned Coal Mine Stores for Colorado Peak-Day Demands," *Pipe Line Industry*, Vol. 49, Sept., 1978, pp 55–58.

Clausing, Russell G., "Gas Storage and Archeology in Historic Areas," *A.G.A. Operating Section Proceedings 1981*, American Gas Association, Arlington, VA, p. T-314.

Coleman, Dennis D., "Gas Identification by Geochemical 'Fingerprinting,'" *A.G.A. Operating Section Proceedings 1987*, American Gas Association, Arlington, VA, p. 574.

Colonna, Jean and Jean-Francois Carriere, "Cushion Gas and Working Volume—For Each Geological Structure, Its Optimal Management," *A.G.A. Operating Section Proceedings 1985*, American Gas Association, Arlington, VA, 1985, p. 665.

Compton, Anthony, "Storage, Its Economics and Role in a Changing Energy Environment," *A.G.A. Operating Section Proceedings 1981*, American Gas Association, Arlington, VA, p. T-291.

Cowart, Vicki, "Borehole Seismic Technology and Applications to Gas Storage," *A.G.A. Operating Section Proceedings 1990*, American Gas Association, Arlington, VA 1990, p. 612.

Crow, C.V., "A Review of Problems and Their Solutions in the Hillsboro (IL) Gas Storage Field," *A.G.A. Operating Section Proceedings 1982*, American Gas Association, Arlington, VA, p. T-431.

Fay, W.C. and P. Eng, "Modeling Seasonal Storage Operations," *A.G.A. Operating Section Proceedings 1986*, American Gas Association, Arlington, VA, 1986. p. 793.

Foley, Dwayne L., "The Role of Storage in Unbundled Services," *A.G.A. Operating Section Proceedings 1988*, American Gas Association, Arlington, VA, 1988. p. 552.

Ford, W.G.F., "Foamed Acid Stimulation: An Effective Stimulation Fluid," *A.G.A. Operating Section Proceedings 1981*, American Gas Association, Arlington, VA. p. T-317.

Gentges, Richard J., "Monitoring Casing Corrosion in Natural Gas Storage Fields," *A.G.A. Operating Section Proceedings 1985*, American Gas Association, Arlington, VA, 1985, p. 684.

Gladish, Lowell E., "Citizens Gas and Coke Utility's Experience on Short-Radius Horizontal Directional Drilling," *A.G.A. Operating Section Proceedings 1989*, Arlington, VA, 1989, p. 575.

Gude, A.J., III, and F.A. McKeown, 1952, *Results of Exploration at the Old Leyden Coal Mine, Jefferson County, Colorado*: U.S. Geological Survey, TEM Report 292, p. 13.

Haddenhorst, H., "Underground Storage of Natural Gas in West Germany," *A.G.A. Operating Section Proceedings 1983*, American Gas Association, Arlington, VA, 1983, p. T-344.

Hardgrave, John J., "Bethel Gas Storage Cavern 2-A Downhole Casing Inspection Project," *A.G.A. Operating Section Proceedings 1985*, American Gas Association, Arlington, VA, 1985, p. 672.

Hardy, H. Reginald, Jr., "Microseismic Monitoring of the New Haven Storage Reservoir," *A.G.A. Operating Section Proceedings 1981*, American Gas Association, Arlington, VA, 1981. p T-296.

Hardy, H. Reginald, Jr., "Rock Mechanics Aspects of the Design of Salt Caverns for the Storage of Natural Gas," *A.G.A. Operating Section Proceedings 1982*, American Gas Association, Arlington, VA, 1982, p. T-409.

Hardy, H. Reginald, Jr., "Feasibility of Utilizing Microseismic Techniques for the Evaluation of Underground Gas Storage Reservoir Stability," *Monograph on Project 12-43 of the Pipeline Research Com-*

mittee of the American Gas Association at the Pennsylvania State University, American Gas Association, Arlington, VA 1975.

Hermann, Tim and Tim Koch, "HDPE Instrumentation at Sayre Storage A Case History, " *A.G.A. Operating Section Proceedings 1988,* American Gas Association, Arlington, VA, 1988, p. 575.

Hirsh, James. E., "Gravel-Packing High-Deliverability Gas Storage Wells at the McDonald Island Gas Storage Field: A Case Study, " *A.G.A. Operating Section Proceedings 1986,* American Gas Association, Arlington, VA, 1986, p. 782.

Hoffstetler, Floyd P., "Storage Field Internal Corrosion Investigation and Rehabilitation," *A.G.A. Operating Section Proceedings 1984,* American Gas Association, Arlington, VA, 1984, p. 182.

Hoglund, Paul A., "A Simple Model of an Underground Storage Field," *A.G.A. Operating Section Proceedings 1986,* American Gas Association, Arlington, VA, 1986, p. 829.

Johnson, Winston A. II and William H. Healy, Jr., "Design of the Washington Ranch Storage Project," *A.G.A. Operating Section Proceedings 1983,* American Gas Association, Arlington, VA, 1983, p. T-329.

Kannberg, L.D. and R.D. Allen, "Recent Advances in Compressed Air Storage in Underground Reservoirs," *A.G.A. Operating Section Proceedings 1984,* American Gas Association, Arlington, VA, 1984, p. 383.

Katz, Donald L., and R.L. Lee, *Natural Gas Engineering Production and Storage,* New York, McGraw Hill, 1990, pp 760.

Katz, Donald L., "Handling Verifications in Aquifer Pressure in Gas Storage, *A.G.A. Operating Section Proceedings 1985,* American Gas Association, Arlington, VA, 1985, p. 675.

Katz, Donald L. and Frederick H. Clark, "Finders Keepers," *A.G.A. Operating Section Proceedings 1985,* American Gas Association, Arlington, VA, 1985, p. 688.

Katz, *Flow of Gas From Reservoirs,* University of Michigan, Ann Arbor, 1955.

Katz, Donald L., "Underground Storage of Natural Gas," *Handbook of Natural Gas Engineering,* New York, McGraw Hill, 1959, pp 655–695.

Katz, Donald L., et al, "Effect of Unsteady-State Aquifer Motion on the Size of an Adjacent Gas Storage Reservoir," *Journal of Petroleum Technology,* Vol. 11, Feb., 1959, pp 18–22.

Katz, Donald L., "Aquifer Behavior in Gas Production," *Oil and Gas Compact Bulletin,* Vol. 19, June 1960, pp 14–20.

Katz, Donald L., M. Rasin Tek, Keith H. Coats, Marvin L. Katz, Stanley C. Jones, "Movement of Underground Water in Contact with Natural

Gas," *Monograph Prepared on Project NO-31 at the University of Michigan for the Pipeline Research Committee of the American Gas Association,* American Gas Association, New York, Feb. 1963.

Katz, Donald L., and M.R. Tek, "Threshold Pressure Phenomena in Porous Media," *Proceedings of 42nd Annual SPE of AIME Fall Meeting, Oct. 1–4, 1967,* 1967. (Reprint No. SPE-1816) 12 pp

Katz, Donald L. and K.H. Coats, *Underground Storage of Fluids.* Ann Arbor, MI, Ulrich's Books, 1968. pp 63–66.

Katz, Donald L., "Gas Penetration into Deeper Strata in Aquifer Storage," *Preprints of 44th Annual SPE of AIME, Fall Meeting, Sept. 28–Oct. 1, 1969.* Preprint No. SPE 2561, 12pp.

Katz, Donald L. and M.R. Tek, "Storage of Natural Gas in Saline Aquifers," *Proceedings of the 50th Annual AGU Meting, Apr. 21–25, 1969.* Trans. Paper No. H69.

Katz, Donald L., "Monitoring Gas Storage Reservoirs," *Preprints of AIME Midwest Oil and Gas Industry Symposium,* 1971. Preprint No. SPE-32–87, 8 pp.

Katz, Donald L., "Outlook for Underground Storage," *Proceedings of 4th Northern Ohio Geological Society Salt Symposium,* Vol. 1, 1974, pp 253–258.

Katz, Donald L., et al, "Application of Radial Simulation Model," *Society of Petroleum Engineers Preprints of 51st Annual Fall Meeting, New Orleans, 1976.* SPE Paper No. 5883. Dallas, 1976.

Katz, Donald L. and V. Bagrodia, "Gas Migration by Diffusion in Aquifer Storage," *Journal of Petroleum Technology,* Vol. 29, Feb. 1977, pp 121–122. (SPE-5882).

Katz, Donald L., "Containing Natural Gas in Underground Storage Fields, Part 1," *Pipe Line Industry,* Vol. 49, Sept. 1978, pp 62–64.

Katz, Donald L., "Containing Natural Gas in Underground Storage Fields, Part 2," *Pipe Line Industry,* Vol. 49, Oct., 1978, pp 71–74.

Katz, Donald. L., "A Look Ahead in Gas Storage Technology," *A.G.A. Operating Section Proceedings 1981,* American Gas Association, Arlington, VA, 1981. p T-283.

Kelly, R.E., "Cavern Storage of Natural Gas Using an Abandoned Coal Mine in Colorado," *Proceedings of the American Gas Association, Operating Section 1961,* American Gas Association, New York 1961.

Kelly, R.E., "Storing Gas in an Abandoned Mine," *Pipeline Industry,* Vol. 15, No. 1, July 1961, pp 46–51

Kothari, K.M., J.L. Bashbush, A.D. Modine, and J.E. McElhiney, "Storage Reservoir Modeling," *A.G.A. Operating Section Proceedings 1988,* American Gas Association, Arlington, VA, 1988, p. 561.

Kraus, Michael, "Extinguishing of Natural Gas Fires," *A.G.A. Operating*

Section Proceedings 1984, American Gas Association, Arlington, VA, 1984, p. 62,

Kresse, Thomas J., "CO_2 Removal Prior to Natural Gas Storage Injection," *A.G.A. Operating Section Proceedings 1987*, American Gas Association, Arlington, VA, 1987, p. 562.

Legatski, M.W., and D.L. Katz, "Dispersion Coefficients for Gases Flowing in Consolidated Porous Media," *Society of Petroleum Engineers Journal*, Vol. 7, No. 1, March, 1967, pp 43–53.

MacDonald, Robert C., "Assessing the Value of Observation Well Data in Gas Storage Inventory Monitoring," *A.G.A. Operating Section Proceedings 1985*, American Gas Association, Arlington, VA, 1985, p. 710.

Mato, Stephen A., Jr., "Multichannel Casing Inspection Instrument," *A.G.A. Operating Section Proceedings 1987*, American Gas Association, Arlington, VA, 1987, p. 577.

Meddles, R.M., 1978, "Underground Storage in the Leyden Lignite Mine," *Energy Resources of the Denver Basin, Rocky Mountain Association of Geologists Fall Field Conference, Colorado Springs, CO, October 9, 1978*, Denver, 1978 (Conf. No. 78 06195)

Morrow, Richard E., "Operating Gas Storage Fields in Urban Environments," *A.G.A. Operating Section Proceedings 1988*, American Gas Association, Arlington, VA, 1988, p. 556.

Pirkle, Robert J., "Near-Surface Geochemical Monitoring of Underground Gas Storage Facilities," *A.G.A. Operating Section Proceedings 1986*, American Gas Association, Arlington, VA, 1986, p. 804.

Pope, Daniel H., David M. Dziewulski, Timothy P. Zintel, Henry Aldrich and James. R. Frank, "Microbiologically Influenced Corrosion and Hydrogen Sulfide Production in Gas Industry Facilities," *A.G.A. Operation Section Proceedings 1989*, American Gas Association, Arlington, VA, 1989, p. 578.

Roberts, Joseph L., Edward Dereniewski and Yusuf Hekim, "Deliverability Interference in Gas Storage Reservoirs," *A.G.A. Operating Section Proceedings 1982*, American Gas Association, Arlington, VA, p. T-403.

Rose, J.W., " Storage Field Automation: A Case Study of the Lyons, Kansas, Underground Storage Field," *A.G.A. Operating Section Proceedings 1981*, American Gas Association, Arlington, VA, 1981. p T-274.

Rzepczynski, Walter M., "Foreign Technology Regarding Underground Natural Gas Storage," *A.G.A. Operating Section Proceedings 1988*, *A.G.A. Operating Section Proceedings 1989*, American Gas Association, Arlington, VA, 1988, p. 570.

Smith, Jeffrey Duff, "Lateral Drilling Technology: Short-Radius Technique," *A.G.A. Operating Section Proceedings 1987*, American Gas Association, Arlington, VA, 1987, p. 585.

Swanson, R.K., M. Kilman, Armando De Los Santos, and David Miller, "Corrosion Damage Assessment in Gas Storage Well Casing by Measurement of Advanced Magnetic Perturbation Logging," *A.G.A. Operating Section Proceedings 1987*, American Gas Association, Arlington, VA, 1987, p. 550.

Tek, M. Rasin, E.O. Udegbunam, S. Schwartz, R.L. Lee, R.W. Logan, and J.S. Ahluwalia, "Inventory – Migration – Deliverability in Underground Storage," *Monograph Prepared on Project PR-26-133 at the University of Michigan for the Pipeline Research Committee of the American Gas Association*, American Gas Association, Arlington, VA, 1986.

Tek, M. Rasin, J.O. Wilkes, "New Concepts in Underground Storage of Natural Gas," *Monograph on Project PO-50 at the University of Michigan for the Pipeline Research Committee of the American Gas Association*, American Gas Association, New York, March, 1966.

Thompson, George W., "Role of the Landman in the Acquisition of Underground Storage Rights," *A.G.A. Operating Section Proceedings 1984*, American Gas Association, Arlington, VA, 1984, p. 85.

Thompson, George W., "Role of the Landman in the Acquisition of Underground Storage Rights," *A.G.A. Operating Section Proceedings 1984*, American Gas Association, Arlington, VA, 1984, p. 85

Tillman, Ned "Soil Gas Surveying," *A.G.A. Operating Section Proceedings 1989*, American Gas Association, Arlington, VA, 1989, p. 607

Tutt, Charles, Jr., "Natural Gas Storage Field Inventory Verification by Statisical Analysis and Observation Well Pressures," *A.G.A. Operating Section Proceedings 1984*, American Gas Association, Arlington, VA, 1984, p. 516

Uding, Bill, "Storage Reservoir Modeling Using a PC," *A.G.A. Operating Section Proceedings 1989*, American Gas Association, Arlington, VA, 1989, p. 597.

Walbe, Kim, "Utilization of the Borehole Television Camera in Open and Cased Holes, *A.G.A. Operating Section Proceedings 1989*, American Gas Association, Arlington, VA, 1989, p. 605

Weideman, O. B., "An Understanding of Underground Gas Storage," *A.G.A. Operating Section Proceedings 1980*, American Gas Association, Arlington, VA, 1980, p. D-291.

Well, John A., John B. Agnew and Ron E. Coker, "Development and Use of an Operating Model of the Worsham Steed Gas Storage

Field," *A.G.A. Operating Section Proceedings 1985,* American Gas Association, Arlington, VA, 1985, p. 699.

Weible, Rudy, "Gas Storage and Enhanced Oil Recovery," *A.G.A. Operating Section Proceedings 1990,* American Gas Association, Arlington, VA, 1990, p. 625.

Whims, Michael J., "Gas Storage in Northern Michigan's Gas-Condensate Reefs: An Update," *A.G.A. Operating Section Proceedings 1981,* American Gas Association, Arlington, VA, 1981. p T-280.

Witherspoon, P. A., J. C. Radke, Y. Shikari, K. Pruess, P. Persoff, S. M. Benson and Y. S. Yu, "Feasibility Analysis and Development of a Foam-Protected Underground Natural Gas Storage Facility," *A.G.A. Operating Section Proceedings 1987,* American Gas Association, Arlington, VA, 1987, p. 539.

Witherspoon, P. A., M. S. King and C. J. Radke, "Foam-Protected Natural Gas Storage Reservoirs," *A.G.A. Operating Section Proceedings 1983,* Arlington, VA, 1983, p. T-350.

Index

A
abandoned coal mines, *see also*
    caverns, mined, 7
  Lyden mine storage facility, 77
absolute open flow, 20
Aliso Canyon storage field, 165
anticline, 5
aquifers, 5, 41
  Herscher storage field, 41
  pressure, radius, time
    equation, 48
  sands, 45
  water movement, 48, 50

B
base gas, 12, 18
basement rock, 12
Bistineau storage field, 117,
    118, 120-12, 134
blanket, 71

C
caprock, aquifer, 5, 53
casing, 26
  joint strength, 27
  minimum yield strength, 26
  pipe body yield strength, 26
  yield strength collapse, 27
casing string, minimum yield
    strength, 26
caverns, leached, 6

  depth, 74
  volume, 63
caverns, mined, 7
coal mines, *see* caverns, mined
coke bottle cavern, 71
compressibility, 36
  aquifer, 46
compressibility factor, 36
cores, 30
cushion gas, 52, 73
  aquifer, 52

D
deliverability, 10, 20
  field, 23
  leached cavern equation, 65

E
electric logs, 31

F
formation pressure equation, 20
fracture-connected cavern, 69

G
gas bubble, 5, 58
gas displacement, 30
gas in place equation, 14, 17
gas wafer, 58

H
Herscher storage field, 41
Horner, 47